Sensing the Perfect Tomato

An Internet of Sensing Approach

Sensing the
Perfect Tomato
An Internet of Sensing Approach

Denise Wilson

CRC Press
Taylor & Francis Group
Boca Raton London New York

CRC Press is an imprint of the
Taylor & Francis Group, an **informa** business

CRC Press
Taylor & Francis Group
6000 Broken Sound Parkway NW, Suite 300
Boca Raton, FL 33487-2742

First issued in paperback 2021

© 2019 by Taylor & Francis Group, LLC
CRC Press is an imprint of Taylor & Francis Group, an Informa business

No claim to original U.S. Government works

ISBN-13: 978-0-367-08676-3 (hbk)
ISBN-13: 978-1-03-209292-8 (pbk)

Library of Congress Cataloging-in-Publication Data

Names: Wilson, Denise (University of Washington), author.
Title: Sensing the perfect tomato : an Internet of sensing approach / author: Denise Wilson.
Description: Boca Raton, FL : CRC Press, Taylor & Francis Group, 2019.
Identifiers: LCCN 2019000729| ISBN 9780367086763 (hardback : alk. paper) | ISBN 9780429023729 (ebook)
Subjects: LCSH: Tomatoes. | Tomatoes--Remote sensing. | Internet of things.
Classification: LCC SB349 .W46 2019 | DDC 635/.642--dc23
LC record available at https://lccn.loc.gov/2019000729

Visit the Taylor & Francis Web site at
http://www.taylorandfrancis.com

and the CRC Press Web site at
http://www.crcpress.com

To my husband Barry, my sister Heidi, and good coffee.
Without them, no chapter would be complete.

Contents

Author

Denise Wilson is a professor in the department of electrical and computer engineering and adjunct professor in the school of environmental and forest sciences at the University of Washington in Seattle. She is also founder and managing director of Coming Alongside, an environmental services nonprofit organization whose mission is to make hazards posed by the environment to human and animal health visible and actionable. She received a BS degree in mechanical engineering from Stanford University (1988), MS and PhD degrees in electrical engineering from the Georgia Institute of Technology, Atlanta, in 1989 and 1995, respectively, and an MEd from the University of Washington in 2008. Denise has published over 40 articles in peer-reviewed journals and over 100 articles in peer-reviewed conferences on topics ranging from sensors/sensing systems to analog circuits to environmental health. She has also published three book chapters and developed extensive web-based educational materials on environmental health and the environmental impacts of technology (labs.ee.washington.edu/community and www.comingalongside.org).

CHAPTER 1

Tomatoes in History

Fʀᴇsʜ? Jᴜɪᴄᴇ? Pᴀsᴛᴇ? Cᴀɴɴᴇᴅ? Stewed? Diced? Pureed? Sun Dried? Salsa? Sauce? Ketchup?

With many varieties and varied uses, the tomato is one of the world's most popular fruits, despite being hopelessly inept as a biofuel feedstock. For the foreseeable future, the tomato is likely to continue proliferating throughout kitchens and across dining tables around the world, while remaining far afield from vehicle fuel tanks. Profits from tomatoes are likely to continue to climb as well because among high-value crops, tomatoes are leaders among those crops that are legal, yielding over a million dollars in sales per square kilometer of crop. While these numbers may seem to be nothing compared to the multiple millions of dollars reaped per square kilometer by harvesting opium, coca, or cannabis, the savings in legal fees and jail time can easily justify choosing the tomato over these other even more highly valued crops (Desjardins 2014, FAOSTAT 2018, UNODC 2014).

Exported tomatoes alone accounted for over 8 billion dollars in sales in 2017 (OEC n.d.). Combining exports and domestic use, 340 billion pounds (170 million tons) of tomatoes are grown for fresh consumption and processed tomato products around the world every year, representing over 50% growth in the early part of the twenty-first century (FAOSTAT 2018). The area it takes to facilitate such massive production is over 11 million acres. China is the world's leader in producing tomatoes, with India, the United States, the European Union, and Turkey filling in the top five to the tune of 70% of global production (FAOSTAT 2018). The United States produces 35 billion pounds of tomatoes every year, and California and Florida lead the pack with about two-thirds of the total (USDA 2018). Furthermore, California produces over 90% of those tomatoes in the United States used for processed tomato products (FREP 2013). Tomato sales just keep multiplying as demand continues to increase.

Consumers in the United States have a love affair with tomatoes that continues to march ever upward with consumption of fresh tomatoes and tomato products exceeding 10 billion pounds per year (USDA 2008). Historically, though, this love of tomatoes was slow to blossom. In fact, while tomatoes originated in South America, they arrived in the United States rather circuitously by way of Europe after spending decades traveling the globe and becoming an integral part of many regional diets, cultures, and cuisines.

THE ORIGIN

Tomatoes originated from and grow wild at moderate altitudes in the Andes Mountains of Peru, Ecuador, and Bolivia in South America. The moderate temperatures, rainfall, and climate in their native habitat underlie why many cultivated tomatoes in the modern world are particularly sensitive to chilly temperatures and to very hot or very dry weather. Native peoples of South America are likely to have cultivated tomatoes for the first time, but it was centuries later that the tomato finally made it to Mexico where it was formally domesticated. Some think that seeds from the cherry tomato flew by way of a strong wind from South America to Mexico. Others think that the tomato made its way to Mexico through Indian migration, similarly to how Andean maize arrived in Mexico. But regardless of how they got there, tomatoes were among the last in line to be domesticated, largely because they are so perishable and vulnerable to bruising and damage. And that is how the migrated tomato languished, waiting hundreds of years until the white man arriving into Mexico from Spain and Portugal transported the first tomatoes and seeds back to Europe. As evidenced by names like pomi d'oro (Italian for gold apple), the tomato at the time was likely to have been clothed not in modern shades of red but rather golden shades of yellow. Upon its arrival in Europe, the tomato was welcomed neither with open arms nor open mouths. Instead, it was shunned and dismissed as mere poison.

THE BAD

The hundreds of years that it took the tomato to tour the world and finally enter into the colonial United States can be largely attributed to the British belief at the time that tomatoes were poisonous. This belief stemmed from the tomato's membership in the nightshade family, whose most famous member at the time was the belladonna plant. Not only was belladonna believed to have put Juliet into her deep sleep, it was also held responsible for poisoning a sizeable number of armies, enemies, and hapless targets throughout history. Belladonna, adorned with her toxic alkaloids of atropine, hyocyamine, and scopolamine, remains responsible for isolated cases of poisoning even in modern times, especially in children who consume its berries without thought to dire consequences (Berdai et al. 2012). Because the tomato is related to the belladonna plant, it was also assumed to be toxic. Unbeknownst to those who assumed the worst, however, the tomato contains only enough solanine in its leaves and immature fruit to induce mild gastrointestinal distress and vomiting, but nothing nearly so disastrous as deep sleep or death (Barceloux 2009). Europeans didn't figure this out for some time and convictions proliferated with regard to the dark side of the tomato. These rumors were only made worse by the fact that the tomato's acidity drew lead from pewter plates, which ultimately ended up in the stomachs of those who ate from pewter wares. Subsequent upticks in illness and death among the aristocratic who had a preference for pewter were blamed, logically or not, on the tomato (Smith 2013).

Perceptions of poison were not the only reason that tomatoes did not fare well in many countries. For example, in Italy, the fact that tomatoes grew close to the ground diminished the fruit's social status to the point that eating tomatoes was beneath most, including many peasants (Gentilcore 2010). Furthermore, tomatoes were often confused with another member of the nightshade family, the mandrake, which has a long history as a fertility drug and aphrodisiac. As a result, Puritans were rumored to shy away from the tomato in order to reduce temptation and maintain spiritual purity (Smith 2013). Also, the unique flavor of the tomato did not help its uphill battle to be included in European and early American cuisine. Tomatoes are often an acquired taste and not one that is necessarily appealing at first bite. The odd flavors of tomatoes often led to their dismissal as "sour trash" (LeHoullier 2014), which further delayed their entry into mainstream cuisine. The list of reasons used to reject the tomato as an edible was so long that it stayed off the dinner plate and was confined to very limited use and equally limited popularity for many years. Those many years stretched into almost two centuries during which the tomato remained "toxic," erotic, inferior, and sour. But, could the poor tomato remain down and out forever, or would it emerge victorious over all who had shunned it? Answering this question requires taking a few more steps forward in history.

THE PRETTY

The tomato suffered its ill-deserved but widespread reputation as a questionable edible in Europe and North America for multiple centuries after Columbus, Cortez, and comparable conquistadors arrived back in Europe bearing the first fruits from their travels. Even though it did not grace the dinner plate or the salad bowl, the tomato did often land in the centerpiece of fine dining. Italian nobility in particular are rumored to have been enchanted by the aesthetics of the little golden apple and incorporated it into many a tabletop decoration. As a result, the pressure to craft the aesthetics of the cherished tomato into a wide range of shapes and sizes mounted, and throughout much of the seventeenth and eighteenth centuries, Italian scientists took on the challenge to produce as many decorative varieties as possible. The British did likewise, growing the tomato as an ornamental plant and using its fruit for many a decoration outside of the garden. But, the persistence of both the Italians and the Spanish in selectively breeding the tomato is what eventually led to the fruit's emergence from the art world to the edible world, somewhere around the mid-eighteenth century. The golden apple (pomodoro) had graduated into a plethora of colors, shapes, and sizes that were ready for something more than decoration (Vegetable Facts 2018).

THE TURNING POINT

Once notions of toxicity had passed, the tomato quickly evolved from a prized decoration to a highly valued fruit. In Spain and parts of Italy controlled by Spain during the seventeenth century, an early ratatouille was born consisting of fried tomatoes, onions, squash, and eggplant. From ratatouille, tomato-based recipes proliferated, assisted by the ability of the flavorful tomato to substitute for expensive spices at a mere fraction of their cost. Margherita pizza, pasta pomodoro, and a myriad of other tomato-infused meals grew in number and popularity until tomato-based recipes were so widespread that many Europeans soon assumed that tomatoes had originated in Italy and their culinary history there must have spanned many centuries' worth of refinement in breeding and cooking (Grand Voyage Italy 2016).

In North America, the turning point for the tomato, while equally dramatic, was quite different. In a young United States, a story that is as popular as it is unproven maintains that a single horticulturist transformed the attitude of early Americans toward the hapless, toxic, sinfully tempting, and woefully sour trash tomato. The horticulturist was a man named Robert Gibbon

Johnson who, in 1820, made a big production of climbing the steps of the courthouse in Salem, Massachusetts, and eating a large volume of tomatoes in one sitting. Purportedly, once Mr. Johnson had clearly survived the ingestion of these devilish fruits with no ill effects, Americans began coaxing the tomato out of the shadows and onto the menu in earnest. Similar storytelling around other individuals is found in early American historical lore and many may be more the fodder of legend than fact. While the real story behind the tomato turning point in American history may never be fully known, it is clear that in the early 1800s, something did indeed happen to transform the American attitude toward the tomato and move the people forward in adopting it as an acceptable edible (Smith 2001).

THE CRAZE

While colonial Americans were busy trying to figure out how to grow more and more fruitful tomato plants, immigrants from Italy and other European countries were busy importing canned varieties from their homeland. By the end of the nineteenth century, Italians used such large quantities of tomatoes in their cooking that mass canning of tomatoes was born and large quantities of canned peeled tomatoes were being exported to the United States to support demand from Italian immigrants (Grand Voyage Italy 2016). At the same time, an American, Alexander W. Livingston, had developed multiple tomato varieties that could grow in almost every state of the union in gardens ranging from the residential backyard to large commercial farms (Yearbook of Agriculture 1937). The goal for commercial crops was a tomato that was smooth, uniform, sweet, and colorful, which led to the Paragon variety of tomato that continues to thrive in modern tomato farming. Whether canned and imported or fresh and domestic, tomatoes proliferated throughout the American states, and by 1920, Americans were eating 27 pounds of canned tomato products and over 14 pounds of fresh tomatoes (not including homegrown crops) every year (Bezilla n.d.).

But, the true testament to the exploding popularity of the tomato came in 1893. Despite clearly being a seed-bearing fruit that emerges from the ovary of a flowering plant as all fruits do, the tomato was declared by the Supreme Court to be a vegetable despite clear scientific evidence to the contrary. With such a declaration, tomatoes could henceforth be taxed according to the Tariff Act of 1883. The rationale? "Do you eat it for dessert? Fruit. Do you eat it for dinner? Vegetable. Problem Solved"—Supreme Court Justice Horace Gray (Eschner 2017).

THE BOOM

Despite its high value and notwithstanding the taxation, not all was rosy on the tomato farm. Unlike many other high-volume crops like corn and soybean that had advanced into industrialized and mechanized agriculture, tomatoes remained too fragile for automated handling and harvesting. California, with its massive tomato acreage, wasn't going to allow such manual handling shenanigans to go on forever. From the University of California at Davis, a seed specialist (Jack Hanna) and aeronautical engineer (Colby Lorenzen) joined forces to pursue the mechanization of tomato farming with such intensity and single mindedness that they withstood over a decade of repeated failures before becoming successful. By 1965, the massive $200,000 UC-Blackwelder tomato harvester had gained a reputation for breaking down more often than it ran, but a few years later, it had transformed the California tomato industry by mechanizing the harvesting of tomatoes that were headed for processing facilities (Boom California 2013). While the mechanization displaced thousands of workers and put many small and medium-sized farms out of business, it enabled California to become a tomato powerhouse, producing over 11 million tons of processing tomatoes annually at over 50 tons per acre of land (USDA ERS 2017). Today, over 90% of processing tomatoes produced in the United States originate on farms in California (Hartz et al. 2008). Successful as this technology has been for mass mechanization and preparation of processing tomatoes, a counterpart technology for handling and harvesting tomatoes for fresh consumption has not yet arrived in the mainstream. As a result, tomatoes are one of the few remaining crops of high global volume that must still be harvested by hand if they are to grace the salad bowl rather than the ketchup bottle. As long as manual handling and harvesting remains the norm, the technologies used to support improved growing and farming practice for fresh tomatoes are likely to look far different than for other high-volume crops around the world.

THE UNTHINKABLE

It is hard to meet a tomato seedling that won't someday produce tomatoes. Planted in the summer almost anywhere in the United States, a tomato seedling will find a way to produce tomatoes. Perhaps not the most or the prettiest, but the seedling will grow, mature, and produce nevertheless in all states in the union, with the exception of one state: Florida.

But, for some unthinkable and myopically economic reason, a large percentage of fresh tomatoes in the United States are indeed grown in the Sunshine State, where the soil and climate are outrageously inhospitable. In 2009, Florida produced a little over 800 million pounds of the 2.7 billion pounds of tomatoes grown in the United States for fresh consumption (Guan et al. 2017). Massive production continues despite the fact that there are few if any logical reasons to grow tomatoes in Florida. Among other things, tomatoes don't like sand, hot weather, or days shorter than 12 hours. But, in the wintertime, tomato plants are placed in exactly these conditions, which abound in the Florida winter. Forcing them to produce in such foreign conditions requires a broad range of pesticides, herbicides, and fertilizers that force the tomato to grow at the cost of contaminating soil and water and damaging critical wetland ecosystems throughout the central and southern part of the state. This massive environmental harm is compounded by the fact that farming tomatoes in Florida has also stimulated the proliferation of illegal trafficking and inhumane treatment of agricultural workers.

In sum, tomatoes, almost singlehandedly, have created an untenable and unsustainable state of agriculture in the Sunshine State. Despite this, tomato production in Florida continues to thrive, reaching over 950 million pounds in 2015 (Guan et al. 2017). Fortunately, organizations have sprung up to protect agricultural workers involved in harvesting tomatoes, but the damage to the environment and ecosystem continues to advance, almost unchecked, to meet the unrealistic expectation of many consumers for a year-round supply of fresh tomatoes (NPR 2011).

While warm-weather states like Florida have produced a volume of tomatoes that any self-respecting tomato plant would have thought unthinkable, extensive open field production in this state and in other warm winter climate states has also prevented the expansion of an industry that may be more lucrative and sustainable—the hothouse or greenhouse approach.

THE HOTHOUSE

In the tireless quest for a flavorful tomato, American consumers have turned to greenhouse-grown tomatoes from Canada. Tomatoes are Canada's largest greenhouse export, and over 99% of its greenhouse crop is exported to the United States (Canada 2017). Vast greenhouses have popped up across Canada, so much so that it now has the highest concentration of greenhouses in North America. All because, in the mid-twentieth century, a few key farmers decided that they no longer wanted to be victims to the cold weather endemic to Canadian agriculture and took a lesson or two from highly successful Dutch farmers in how to grow tomatoes competitively in greenhouse climates. The end result is a tomato that contently grows indoors with its roots immobilized in crushed rock encased in a plastic bag, fed by the horticultural equivalent of an IV, a thin plastic tube that injects water and nutrients. Grown this way, tomato plants produce over 10 times more volume than open-field production of tomatoes. Although greenhouse production costs more, the end product is a more flavorful and appealing tomato than what is grown in many open fields in the United States (Charles 2016). Demand for the colorful, pretty red Canadian greenhouse tomato could well outpace demand for the less flavorful Florida tomato, which will undoubtedly regard its Canadian cousin as a traitor to the cause.

THE FUTURE

Whether drawing from American, Canadian, or Mexican farms, the consumption of fresh tomatoes in the United States has remained steady over the past two decades at around 20 pounds per capita (Statista n.d.). In the same time period, however, total fresh tomato production in the United States has gone from over 4 billion pounds in 2000 to 2.7 billion pounds in 2015 (Guan et al. 2017). Similarly, processing tomato production dropped over 10% from 2015 to 2016 from 29,508 million pounds to 26,379 million pounds (Parr et al. 2018). Competition to meet American tomato demand from Mexico and Canada has played a significant role in these declines (Guan et al. 2017).

Regardless of where the tomatoes are grown, the American demand for tomatoes has come a long way from the "sour trash" days of the seventeenth and eighteenth centuries. In fact, in the early twenty-first century, taste has re-emerged as a driving factor in tomato quality, which in turn has stimulated extensive research that is dedicated to returning flavor to the tasteless red fruit that has become the icon of commercial tomato farming (Handwerk 2017). This renewed interest in flavor has stimulated a whole new perspective on tomato farming. Heirloom tomatoes have emerged into the spotlight at higher prices than conventional varieties and are now a highly lucrative and profitable crop for both small farmers and hobby farmers. Hobby, small, and medium farms now dominate the fresh tomato landscape, some conventional and some organic, but each valuing taste and quality in new and innovative ways.

Expanding and evolving interest in fresh tomatoes has also stimulated a new look at how technology can be used to support the tomato industry. While mechanized harvesters and handling operations came to maturity in the 1960s and 1970s for processing tomatoes, these technologies do not match the needs of smaller tomato farms. Rather, other sensor and support technologies are needed that allow the quality of each individual tomato to be maximized and profits to be achieved via flavor and quality rather than sheer volume. Until ground robots can successfully navigate unstructured environments like those found in open fields and crop rows, however, harvesting tomatoes for fresh consumption is likely to remain dominated by manual harvesting and handling practice. Thus, the role of technology in supporting these farming practices will be to supplement rather than replace what farmers, pickers, and growers already do. Sensor technologies are well positioned to support growing and harvesting tomatoes of greater flavor, improved yields, and higher profit margins. Delivering the perfect, tastiest tomato certainly won't be easy, but it deserves to be the next milestone in the long and complex history of the tomato.

CHAPTER 2

Tomatoes in the Diet

TOMATOES HAVE GROWN A dual personality since emerging from their original reputation as a toxic and unpalatable member of the nightshade family. On the science side of their personality, tomatoes are a *fruit* because they contain seeds and grow from the ovary of a flowering plant. However, to chefs (and the Supreme Court of the United States, for that matter), the tomato remains forever a *vegetable* because it is used almost exclusively in savory rather than sweet dishes. Regardless of which personality the tomato shows, whether fruit or vegetable, few will dispute that the tomato is a healthy addition to the diet (Figure 2.1). In fact, the consumption of tomatoes and olive oil is considered one of the primary reasons individuals living in the Mediterranean region show decreased incidences of coronary heart disease and multiple types of cancers. Eating tomatoes has also been implicated in reduced mortality in children; reduced incidence of respiratory infections; reduced muscular degeneration in the elderly; and reduced risk for oral, pharyngeal, esophageal, prostate, colorectal, breast, and ovarian cancers (Bhowmik et al. 2012, Blum et al. 2005).

It is impossible to distill the health benefits of eating tomatoes down to a single nutrient. This is true of most fruits and vegetables. However, the healthy tomato is perhaps best recognized for the

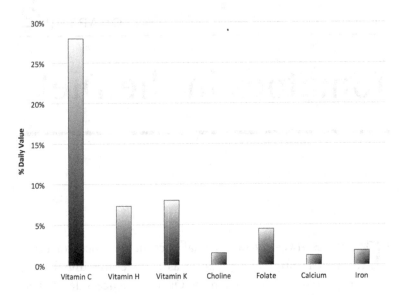

FIGURE 2.1 Nutritional value of the tomato. Although the tomato is best known for being rich in lycopene, it also contains significant amounts of other nutrients.

lycopene it contains. Lycopene ($C_{40}H_{56}$) is an organized string of eight isoprene units, a carotenoid pigment that is responsible for the bright red color of many tomato varieties.

Even more exciting than the bouquet of red that it generously gives to both garden and farm alike, lycopene can decrease the risk of a wide range of chronic diseases and cancers. While lycopene dominates the carotenoid content of the tomato, the fruit also contains three other major carotenoids: α-carotene, β-carotene, and lutein. These phytochemicals all have synergistic and positive effects on human health. And, to compliment tomatoes further, the fruit has the unusual property that lycopene content and nutrient absorption by the body improve with cooking. In contrast, many fruits and vegetables lose substantial nutritional content through cooking or other heat-based processing (Bhowmik et al. 2012).

A tomato is far more than the sum of its carotenoids and contains a broad range of nutrients that contribute to its well-deserved reputation as a healthy addition to most diets, Western or otherwise. Table 2.1 provides a summary of this diverse nutrient content for a single serving of fresh tomatoes and a common tomato product (canned tomatoes). The benefits to human health of the four most plentiful nutrients (vitamins C, H, K, and potassium) as a percentage of recommended daily values (DVs) will be reviewed in further detail in this chapter, as will lycopene, which is available in no other food at the levels that the tomato provides.

VITAMIN C (17 mg OR 28% DAILY VALUE IN ONE MEDIUM FRESH TOMATO)

Ascorbic acid, commonly known as vitamin C, is a water-soluble vitamin that supports healthy body functioning along multiple pathways. Many fruits and vegetables contain vitamin C, with oranges/citrus fruit and red peppers leading the pack. A summary of the vitamin C content of select common fruits and vegetables including the tomato is provided in Table 2.2. Less than a cup of tomato juice can provide up to 33 mg of vitamin C, while a single fresh tomato contributes 17 mg to a healthy diet. This may seem anemic compared to a single serving of orange juice at 93 mg, but the vitamin C content of tomatoes is nevertheless on a par with many other common fruits and vegetables.

Perhaps the most interesting characteristic of vitamin C is that it has both an upside and a downside, as both a protector and killer of cells. At high concentration levels, vitamin C can act as an antioxidant, inhibiting oxidation of molecules in the body and preventing the formation of free radicals, which damage DNA, kill cells, lead to the formation of tumors, and cause cancer (Li and Schellhorn 2007). At even higher concentrations, vitamin C can do the opposite by promoting oxidation of molecules, which leads to cell death (cytotoxicity). High doses of vitamin C have been used in this capacity to selectively damage cancer cells, thereby providing a way to potentially target certain cancer cells during treatment of the disease (Kim et al. 2018). Scientific research supports both positive roles for vitamin C: in neutralizing free radicals before

TABLE 2.1   Nutrient Content in Tomatoes

Nutrient		Units	Daily Value[a]	Fresh Tomato (Medium)[b]	Canned Tomatoes (100 g)
Total Energy		kcal		22	16
Dietary Fiber		g		1.5	1.9
Total Fat		g		0.25	0.25
Vitamin	A	mcg RAE	900	52	20
Vitamin	**C**	**mg**	**60**	**17 (28% DV)**	**12.6 (21% DV)**
Vitamin	**H (Biotin)**	**mcg**	**30**	**2.2 (7.3% DV)**	**N/A**
Vitamin	**K**	**mcg**	**120**	**9.7 (8.0% DV)**	**2.6 (2.2% DV)**
Vitamin	α-Carotene	mcg	N/A	124	N/A
Vitamin	β-Carotene	mcg	N/A	552	245
Vitamin	Lutein + zeaxanthin	mcg	N/A	151	78
Vitamin	Choline	mg	550	8.2 (1.5% DV)	6.4 (1.2% DV)
Vitamin	Folate	mcg	400	18 (4.5% DV)	8 (1.2% DV)
Vitamin	**Lycopene**	**mcg**		**3165**	**2537**
Mineral	Calcium	mg	1000	12 (1.2% DV)	33 (3.3% DV)
Mineral	Iron	mg	18	0.33 (1.8% DV)	0.57 (3.2% DV)
Mineral	Manganese	mg	2.3	0.14 (6% DV)	0.068 (3% DV)
Mineral	Magnesium	mg	420	14 (3.3% DV)	10 (2.4% DV)
Mineral	**Potassium**	**mg**	**4700**	**292 (6.2% DV)**	**191 (4.1% DV)**
Mineral	Phosphorous	mg	1250	30 (2.4% DV)	17 (1.4% DV)
Mineral	Zinc	mg	11	0.21 (1.9% DV)	0.12 (1.1% DV)
Flavanone	Naringenin	mg	N/A	0.8	N/A
Flavonol	Quercetin	mg	N/A	0.7	0.5

Source: Adapted from National Institutes of Health (NIH). 2018b. "Vitamin B7 Fact Sheet for Health Professionals." Accessed September 17, 2018. https://ods.od.nih.gov/factsheets/Biotin-HealthProfessional/; US FDA, Food and Drug Administration. n.d. "Vitamins, Minerals." https://www.accessdata.fda.gov/scripts/InteractiveNutritionFactsLabel/factsheets/Vitamin_and_Mineral_Chart.pdf; USDA, United States Department of Agriculture. n.d. "USDA Food Composition Databases." Accessed March 7, 2018. https://ndb.nal.usda.gov/ndb/search/list.

Abbreviations: g: grams; mg: milligrams; mcg: micrograms; RAE: retinol activity equivalent; DV: daily value.

[a] Daily values based on new FDA dietary labeling guidelines for adults that go into effect on July 26, 2020.

[b] Approximately 123 grams.

TABLE 2.2   Vitamin C Content of Common Fruits and Vegetables

Fruit	Vitamin C	Vegetable	Vitamin C
Orange juice (¾ cup)	93 mg	Red pepper, ½ cup	95 mg
Strawberries, ½ cup	49 mg	Broccoli, cooked, ½ cup	51 mg
Tomato juice, ¾ cup	33 mg	Cabbage, cooked, ½ cup	28 mg
Cantaloupe, ½ cup	29 mg	Spinach, cooked, ½ cup	9 mg
Tomato, 1 medium	17 mg	Green peas, cooked ½ cup	8 mg

Source: Adapted from National Institutes of Health (NIH). 2018c. "Vitamin C Fact Sheet for Health Professionals." Accessed September 18, 2018. https://ods.od.nih.gov/factsheets/VitaminC-HealthProfessional/.

DNA can be damaged and before tumor growth is initiated and in behaving like a pro-oxidant so the body's own free radicals can destroy tumors in their early stages (Block 1991).

As noted above, vitamin C has antioxidant functions, and the health benefits of antioxidants are well established in the literature (Naidu 2003). However, studies that have sought to establish a positive relationship between vitamin C and reductions in breast, esophageal, lung, pancreatic, colorectal, prostate, cervical, and ovarian cancer have been largely inconclusive. The one exception is stomach cancer, as vitamin C has been conclusively linked to a reduced risk of this type of cancer (Naidu 2003).

Similar mixed results have emerged from studies of the relationship between vitamin C intake and atherosclerosis (hardening of the arteries) and cardiovascular disease. Although taken as a whole, it is not possible to say that vitamin C prevents these health problems, the mechanisms by which vitamin C acts on lipids in the body suggest that it has a positive role in protecting the heart and arteries from damage and disease (Naidu 2003).

In addition to its antioxidant functions, vitamin C also supports the production of collagen, which is the most abundant protein in the body and is responsible for maintaining healthy skin, bones, teeth, cartilage, tendons, blood vessels, heart valves, spinal discs, and parts of the eye. The production of collagen, as supported by vitamin C, is also an essential part of allowing the body to heal its wounds and regenerate tissue after injury (Naidu 2003). Versatile vitamin C also supports the biosynthesis of muscle carnitine, which is involved in energy production in the body and production of the neurotransmitter dopamine and the hormones oxytocin and vasopressin.

In addition to this wide range of health benefits, the best-known role of vitamin C is as a preventative or cure for the common cold. Unfortunately, vitamin C has never actually been proven to prevent or cure colds, although it does reduce both the severity and the longevity of even the most persistent cold (Naidu 2003).

Although the daily value of vitamin C established by the Food and Drug Administration is 60 mg, there are many situations where increased intake of vitamin C may be appropriate. Vitamin C can be poorly absorbed in the intestine, especially at higher concentrations, and is unstable in the body, requiring higher and more consistent intake than other vitamins (National Institutes of Health [NIH] 2018e). Stress, smoking, and infections also deplete the vitamin C in the body and require greater dietary intake or supplements to maintain healthy levels than for more healthy individuals (Naidu 2003).

VITAMIN K (9.7 mcg OR 8% DAILY VALUE IN ONE MEDIUM FRESH TOMATO)

Like most fruits and vegetables, tomatoes supply vitamin K to the diet as phylloquinone, thereby providing support for blood clotting, bone health, and healthy blood calcium. Among fruits, tomatoes are a good source of vitamin K (Table 2.3), although green vegetables typically provide more significant amounts of phylloquinone.

Osteoporosis, a condition associated with fragile and brittle bones, is typically associated with aging, women, and deficiencies in calcium or vitamin D. Vitamin K is essential to the modification (i.e., carboxylation) of osteocalcin, a protein in bone that is necessary to maintain healthy bone mineral density (Gundberg et al. 2012). Not surprisingly, some research studies have therefore linked vitamin K levels and intake with higher bone mineral

TABLE 2.3   Vitamin K Content of Common Fruits and Vegetables

Fruit	Vitamin K	Vegetable	Vitamin K
Pomegranate juice, ¾ cup	19 mcg	Spinach, raw, 1 cup	530 mcg
Blueberries, raw, ½ cup	14 mcg	Kale, raw, 1 cup	113 mcg
Fresh tomatoes, 1 cup	14 mcg	Broccoli, boiled, ½ cup	110 mcg
Grapes, ½ cup	11 mcg	Edamame, ½ cup	21 mcg

Source: National Institutes of Health (NIH). 2018d. "Vitamin K Fact Sheet for Health Professionals." Accessed September 26, 2018. https://ods.od.nih. gov/factsheets/VitaminK-HealthProfessional/.

density and reduced fractures (Yaegashi et al. 2008). Supplemental vitamin K in the form of menaquinone has also been shown to reduce hip, vertebrae, and other fractures in many but not all studies, as reviewed in Cockayne et al. (2006).

Unfortunately, taken as a whole, studies of vitamin K have been inconsistent, enough so that no solid recommendation can be made about vitamin K intake (whether by diet or by supplement) and bone health. However, it is clear from these studies that any positive impacts on reducing osteoporosis with vitamin K are impossible by relying on dietary sources alone and proving such benefits will require studying vitamin K supplements in combination with zinc, magnesium, calcium, and vitamin D (Cashman 2007). Like many vitamins and other nutrients, vitamin K must act in concert with other nutrients to achieve substantive health benefits, at least with respect to bone health.

Vitamin K has also been studied in its role in coronary heart disease, specifically vascular calcification, where calcium builds up in the arteries and reduces their elasticity. One of the proteins involved in preventing vascular calcification (Matrix Gla-protein, or MGP) is dependent on vitamin K to function properly. Some studies have shown that supplements of menaquinone are associated with reduced calcification in the arteries (Geleijnse et al. 2004), but phylloquinone (the source of vitamin K present in fruits and vegetables) supplements had no significant effect, while other studies have demonstrated that phylloquinone has some minor impacts on slowing down calcification in the arteries (Shea et al. 2009). Although these studies provided no solid evidence for vitamin K reducing coronary heart disease, they are suggestive and merit further research, especially given the 10% of American adults with chronic kidney disease (American Kidney Fund 2015) who are particularly vulnerable to arterial calcification (Schurgers 2013).

Fortunately, vitamin K deficiencies are extremely rare in healthy individuals who consume a reasonable and varied diet. Nevertheless, two groups of people are particularly vulnerable to vitamin K deficiencies. Newborn infants often have vitamin K deficiencies because phylloquinone transports poorly across the placenta to the fetus. For this reason, newborns are often given a single dose of vitamin K at birth to alleviate deficiencies. Adults who have certain gastrointestinal disorders that affect nutrient absorption, such as cystic fibrosis and celiac disease, are also vulnerable to vitamin K deficiencies and benefit from additional vitamin K in the diet and through supplements. On the other extreme, there is no evidence that too much vitamin K in the body is harmful (NIH 2018f).

POTASSIUM (292 mg OR 6.2% DAILY VALUE IN ONE MEDIUM FRESH TOMATO)

Many individuals do not get enough potassium in their daily diet, making it one of the four major shortfall nutrients in the American diet (health.gov 2015) despite the abundance of potassium in many fruits and vegetables. The potassium content of a typical serving of tomatoes relative to typical servings of other common fruits is summarized in Table 2.4.

Potassium can play a major role in reducing hypertension (high blood pressure). High blood pressure has been implicated as a major risk factor in stroke, coronary heart disease, heart failure, chronic kidney disease, and periodontal disease. In a wide range of studies, increasing potassium intake has been shown to significantly reduce both systolic and diastolic blood pressure.

TABLE 2.4   Potassium Content of Common Fruits and Vegetables

Fruit	Potassium	Vegetable	Potassium
Apricots, dried, ½ cup	1101 mg	Acorn Squash, 1 cup	644 mg
Bananas, 1 medium	422 mg	Spinach, Raw, 2 cups	334 mg
Orange juice, 1 cup	496 mg	Broccoli, Cooked, ½ cup	229 mg
Tomatoes, 1 medium	292 mg	Potato, baked	610 mg
Cantaloupe, ½ cup	214 mg	Asparagus, cooked, ½ cup	202 mg

Source: Adapted from National Institutes of Health (NIH). 2018a. "Potassium Fact Sheet for the Health Professional." Accessed August 20, 2018. https:// ods.od.nih.gov/factsheets/Potassium-HealthProfessional/.

The impacts of potassium on reducing blood pressure are so often positive that increasing dietary potassium intake may be the single most important contribution an individual can make to lower the risk of high blood pressure. In fact, some estimate that increasing potassium intake to 3500 mg per day or more can reduce hypertension by 17% in the adult population (Weaver 2013).

In addition to reducing blood pressure, potassium also plays a strong role in reducing age-related bone loss and the risk of osteoporosis when acting in combination with the flavanoids contained in fruits and vegetables. Potassium also reduces calcium concentration in the urine, suggesting that it improves calcium retention in the body. Studies of the impact of potassium on kidney health have suggested that potassium also reduces the occurrence of kidney stones (Weaver 2013).

Although too little potassium is the more likely risk that most adults in Western culture face, potassium at excess levels can also harm health. Excess potassium levels in the body can cause abnormal heart rhythms, and individuals who have poor kidney function, take certain diuretic drugs, or have severe infections are particularly vulnerable to potassium toxicity (UMMS n.d.).

VITAMIN H/B7/BIOTIN (2.2 mcg OR 7.3% DAILY VALUE IN ONE MEDIUM FRESH TOMATO)

The biotin content of food is not available on the USDA database alongside many other nutrients, and biotin deficiencies among

TABLE 2.5   Biotin Content of Common Fruits and Vegetables

Fruit	Biotin	Vegetable	Biotin
Banana (100 g)	4.0 mcg	Cauliflower (100 g)	7.5 mcg
Strawberries (100 g)	4.0 mcg	Sweet corn (100 g)	6.0 mcg
Fresh tomatoes (100 g)	1.8 mcg	Carrots (100 g)	2.7 mcg
Apples (100 g)	0.9 mcg	Spinach (100 g)	3.4 mcg

Source: Adapted from Hoppner, K. et al. 1994. *Food Research International* 27 (5): 495–497; USDA, United States Department of Agriculture. n.d. "USDA Food Composition Databases." Accessed March 7, 2018. https://ndb.nal. usda.gov/ndb/search/list; Watanabe, T. et al. 2014. *Int J Anal Bio-Sci* 2 (4).

adults who consume a normal diet are rare, with only isolated cases reported among pregnant and breastfeeding women, smokers, the elderly, athletes, acne patients, and alcoholic adults (NIH 2018b, Trüeb 2016). Nevertheless, tomatoes are a rich source of biotin (Table 2.5), and emerging evidence suggests that the role of biotin in supporting health and decreasing the risk and impacts of chronic disease is greater than historically believed.

Although hair loss, skin rashes, and brittle nails are associated with biotin deficiency, the reverse has not been proven to be true. Consuming additional biotin in the diet or as supplements has not been conclusively linked to better hair, skin, or nail health. In fact, biotin appears, at first glance, to be a minor actor in the average person's diet, as eating normally protects against deficiency and biotin supplements have no conclusive effect on human health. Further complicating matters, consuming too much biotin can interfere with laboratory tests used to measure thyroid and other hormones, providing inaccurate results and in some cases incorrectly indicating Graves disease or severe hyperthyroidism (Li et al. 2017, NIH 2018b).

Nevertheless, biotin is a water-soluble vitamin that plays a key role in enabling fatty acids, glucose, and amino acids to be metabolized or used by the human body. Biotin also modifies histones, which organize the DNA in cells, and plays a role in regulating genes and cell signaling (NIH 2018b, Pacheco-Alvarez et al. 2002). Biotin's known role in affecting how and how much glucose can be used by the body suggests that biotin may be important in regulating diabetes. Although studies in humans are limited, biotin has been shown to improve glucose tolerance and reduce glucose levels among mice and rats with non–insulin dependent diabetes (Reddi et al. 1988, Zhang et al. 1996). One study of 447 adults with type 2 diabetes did demonstrate that a combination of chromium picolinate/biotin supplements facilitated greater control of blood sugar levels in overweight and obese subjects (Albarracin et al. 2008). This same treatment has been implicated in reducing risk factors of cardiovascular disease among those with type 2 diabetes (Geohas et al. 2007). In some studies, biotin has also been shown to improve lipid (cholesterol) profiles in otherwise healthy individuals (Fernandez-Mejia 2005), and in combination with other B vitamins, supports improved neurological function and may offset cognitive decline among the elderly (Kennedy 2016).

Overall, the B vitamins outside of B9, B12, and B6 have been poorly studied and the comprehensive health impacts of these vitamins alone and in combination with other B vitamins are only beginning to emerge from the research literature.

LYCOPENE (3165 mcg IN ONE MEDIUM FRESH TOMATO)

No discussion of the nutritional value of the tomato would be complete without considering lycopene. More than 80% of lycopene in the diet in the United States comes from tomatoes (Clinton 1998). Multiple research studies suggest that lycopene is the reason that tomatoes are often associated with reduced risk for both cardiovascular disease and cancer (Story et al. 2010). Eating tomatoes, whether fresh or processed, has been linked to reduced risk for prostate cancer (Giovannucci et al. 2002, Giovannucci, n.d.), as well as reduced occurrence of benign prostatic hyperplasia (BPH), the noncancerous overgrowth of the prostate gland that occurs for many men as they age (Kristal et al. 2008). Eating lycopene-rich foods has also been linked to reduced risk of breast cancer (Cui et al. 2008), lung cancer (Gallicchio et al. 2008), gastric noncardia cancer (Nouraie et al. 2005), and ovarian cancer (Jeong et al. 2009). Research targeting colorectal cancer has linked lycopene intake with reduced risk of the disease as well as reduced occurrence of colorectal cancer biomarkers (Vrieling et al. 2007).

Not all studies are as conclusive with regard to lycopene's protective effects against cancers. Some studies have produced no news and are inconclusive on links between lycopene and prostate cancer (Peters et al. 2007), lung cancer (Satia et al. 2009), colorectal cancer (Leung et al. 2008), gastric cancer (Larsson et al. 2007), and ovarian cancer (Zhang et al. 2007). Clearly, more research is needed to iron out these gray areas in the role of lycopene in protecting against cancer.

In addition to a strong preponderance of evidence to suggest that lycopene decreases risk for a number of cancers, increased lycopene in the body has also been associated with reductions in incidence of cardiovascular (heart) disease as well as a reduction in biomarkers for such disease—see Story et al. (2010) for a recent review. Increases in good cholesterol (high-density lipoprotein) and decreases in bad cholesterol (low-density lipoprotein) and triglyceride levels also occur with increased consumption of tomatoes and tomato products (Shen et al. 2007).

The health benefits of tomatoes and lycopene are surprisingly broad and extend beyond reducing cancers and improving heart health. For example, the gums of patients who consume more lycopene are statistically less likely to bleed or develop gingivitis (Chandra et al. 2007). Lycopene also protects against sunburn induced by UV exposure (Stahl et al. 2001), can reduce the effects of asthma (Wood et al. 2008), and is associated with reduced risk of fractures among the elderly (Sahni et al. 2009).

SUMMARY

Few would argue against tomatoes as an important part of a healthy diet. The benefits to health and well-being of eating tomatoes, whether cooked or fresh, are well researched and documented in the peer-reviewed literature. While lycopene is the most distinct and abundant contribution that tomatoes make to a typical diet, tomatoes contain a wide range of additional vitamins, minerals, and other nutrients that also benefit health.

However, not all tomatoes are created equal. The time at which a tomato is harvested affects how much vitamin C and lycopene the fruit contains; the more ripe the tomato, the greater the concentrations of these two nutrients (Duma et al. 2015). Commercial tomato operations often harvest tomatoes at the mature green stage, when many nutrients are at their lowest, and ripen the fruit with ethylene gas, which can detrimentally impact the nutrient value delivered to the end consumer (UGA Extension 2017). Nutrient content also changes with processing (into canned tomatoes, juices, etc.), sometimes for the better and sometimes for the worse. Lycopene content does not change significantly with processing, but concentrations of the other three primary carotenoids (α-carotene, β-carotene, lutein + zeaxanthin) decrease as a result of cooking and processing at elevated temperatures. pH also changes with processing, as tomatoes used for canning are harvested at lower pH (greater acidity) to ensure product safety. In contrast, minerals are relatively unaffected by processing, but vitamin C and other vitamins may increase as a result of concentration effects during processing. And, nutrient values can also fluctuate with variety of color and tomato. For example, red tomatoes contain more lycopene, more vitamin C, and less sodium than their yellow cousins, while yellow tomatoes contain more zinc and more iron (Georgé et al. 2011, USDA n.d.).

By and large, the tomato has come a long way from being wholly shunned as an edible. In addition to its seemingly endless varieties and multiple colors, the tomato is now harvested, processed, cooked, and eaten in many, many different ways and contributes a kaleidoscope of nutrient compositions, flavors, and health benefits to the human diet. With its increasing popularity, however, the tomato must now meet higher and higher expectations from consumers, many of whom may believe that nothing less than the perfect tomato belongs in breakfast, lunch, snack, or dinner.

CHAPTER 3

The Perfect Tomato

W HAT MAKES FOR A perfect tomato? Well, it turns out that the tomato has been poked, prodded, and analyzed through its many years as a favored fruit, revealing many an insight into what makes the perfect tomato.

COLOR

A consumer tends to judge the goodness of a fruit by its color. Bananas are yellow. Oranges are orange. Tomatoes are red. Grass is green, although not a fruit and largely irrelevant to fruit selection and purchasing decisions.

Tomatoes do not first appear red but green and gradually progress through multiple ripening stages to red (or another final color, i.e., anything but green). In chromameter terms, the green to red color index increases through all stages of ripening, and the blue to yellow color index increases until the final stage of ripening, at which it decreases dramatically (Gautier et al. 2008). The color of a ripe tomato is expressed by a large quantity of carotenoids and a small chlorophyll content. The carotenoids are fat-soluble pigments that contribute red, yellow, and orange colors to the ripe tomato with sensitivity to light but not heat. Chlorophylls contribute green color to the tomato, are sensitive to both heat and acid, and play a critical role in plant health by absorbing light to facilitate photosynthesis (Barrett et al. 2010).

As tomatoes mature and ripen from their initial dark-green color to a bright red, chlorophyll content diminishes while carotenoid concentrations increase (Figure 3.1a). The increase in carotenoid concentration is dramatic, up to 50-fold during the ripening process, and the final product, a ripe tomato, is dominated by the carotenoid lycopene, which makes up as much as 90% of the total carotenoids in the fruit. The carotenoids phytoene, phytofluene, zeta-carotene, gamma-carotene, β-carotene, neurosporene, and lutein also contribute to the final mature color of the tomato. In addition to providing color, carotenoids also play a critical role in plant health by contributing to harvesting light, protecting the plant from excessive light, and attracting pollinators. Lycopene also has many nutritional benefits (Chapter 2), as do the other carotenoids through their many antioxidant properties (Bertin and Génard 2018). Thus, choosing a tomato for its color is not as superficial as it sounds, because a tomato that has ripened to deep red (or other bright color) is typically a healthy one.

Over its long and muddled history, however, the tomato has been bred for a range of colors that extends well behind the traditional fire-engine red. On the one hand, the legacy Paragon tomato and over 30 subsequent varieties of tomatoes developed by A.W. Livingston during the early breeding history of tomato varieties in the United States exemplify the bright uniform red that many consumers continue to expect from a truly ripe tomato. But, red is not the only way a tomato appears in a salad. For example, the Early Wonder tomato prefers to finish the ripening process in a deep but uniform pink shade. Other tomatoes, whether cherry, grape, or full size, ripen to deep shades of orange, yellow, purple, or even black. Still others do not ripen into a uniform color at all. For example, Mr. Stripey is a yellow tomato adorned with red streaks when fully ripe, while Black Krim is a deep purple tomato with a green top or shoulder. Big Rainbow is a tricolor variety boasting green shoulders, a yellow body, and red stripes (Rutgers n.d.,

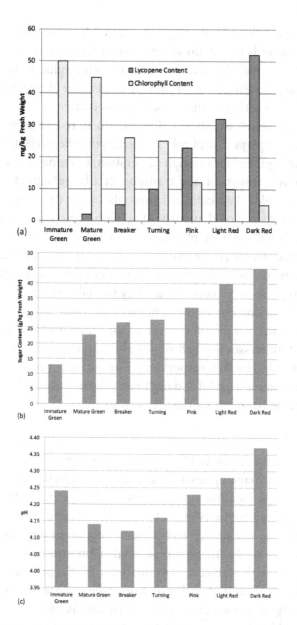

FIGURE 3.1   Properties of tomatoes. During ripening, (a) chlorophyll content decreases, and lycopene increases; (b) sugar content increases; and (c) pH first decreases and then increases again.

Washington State University Extension 2018). The end result is that thousands of tomato varieties are actively cultivated in the world today, and while a majority retain their original uniform red color, an increasing number provide a full range of colors with no shortage of exotic striping and coloration patterns.

Regardless of whether a ripe tomato is red or some other color, color plays a vital role in identifying the perfect tomato. Monitoring the color through maturation, ripening, harvest, and postharvest treatment and transport can be valuable for many reasons. Color data collected throughout the ripening process can scaffold regional models of tomato varieties to facilitate more efficient harvest planning and treatment schedules. For individual farmers, color that is collected by sensors that are more consistent and reliable than the human eye can also identify the optimal point of harvest, particularly for indeterminate varieties that ripen at different times throughout the growing season. And, postharvest, color monitoring can be used to control and optimize ripening processes (e.g., ethylene gas treatment) in a way that is customized to batch transport and distribution times. From farm to table, color is a useful and multifaceted parameter to know and interpret in the quest for the perfect tomato.

FIRMNESS

Firmness and elasticity of a tomato are two parameters among several that make up texture. Texture is also indicated by such properties as mealiness and juiciness (Harker et al. 2002), but firmness and elasticity have been demonstrated to be strong correlates of consumer acceptance and purchasing decisions (Causse et al. 2010). Firmness is also important because it interferes with subjective evaluations of flavor and aroma (Causse et al. 2003), and it predicts shelf life and transport durability (Seymour et al. 2002). Firmness is most commonly measured by the amount of deformation that occurs in response to an applied force. Traditional penetrometers measure the force required to puncture the fruit, while other instruments provide more fine-grained measurements that do not necessarily require damaging the fruit during measurement.

Like color, measuring the firmness of a tomato provides multiple insights into the quality of a tomato, both pre- and postharvest. Before harvest, firmness can indicate internal damage or bruising, optimal time to harvest, and potential negative effects of climate extremes such as extended periods of overnight chilling. After harvest, firmness can be used to assess suitability for transport and sale, affecting yield and economic return. While parameters like mealiness and juiciness are also important to texture, they are harder to measure and often require sampling and processing the tomato to extract these properties. Thus, they are ill suited to the kinds of high-volume measurements that are necessary to deliver consistently fresh tomato products to market.

SUGAR AND ACID CONTENT

Sugar is part of what makes fruit delicious. Tomatoes are no exception. While a tomato is between 90% and 95% water, 50% of what remains is sugar and is represented by glucose, fructose, and sucrose. Sugar content increases during the ripening process (Figure 3.1b), and by the time a tomato is mature, glucose and fructose contribute about equal amounts to the total sugar content, while sucrose is a more minor contributor. Sugar is a major component of flavor both in terms of the total amount of sugar in the fruit and the amount of sugar relative to the acid also present in the fruit (Davies et al. 1981). Sugar content decreases with the size of the fruit, thus in part explaining the tasteless character of many a supersized tomato bred for large-scale farming.

Frequently, overall sugar content in fruit is measured in the field using a Brix meter, where a unit of 1° Brix represents 1 gram of sugar in 100 grams of solution. Brix meters are commercially available and range from inexpensive (less than $20) manually operated devices to fully digitized devices that run into the thousands of dollars. They were originally designed and calibrated to identify sucrose content in solution but are often used to estimate total sugar content by measuring the refractive index of a liquid or solution. Thus, they provide a rough estimate of other sugars in solution, such as the substantial amounts of fructose and glucose that dominate the sugar content in the tomato. Unfortunately, refractive index is not only affected differently by nonsucrose sugars compared to sucrose but is also vulnerable to other changes in parameters that occur naturally during maturation and ripening, such as pH. Thus, using Brix to understand sugar content in tomatoes will always generate, to some extent, fundamental inaccuracies and vulnerability to interference. Differentiating sucrose from fructose and glucose as well as other common interferents often requires more complex laboratory equipment and is frequently not cost effective.

Degradation of tomato acids during maturation and ripening leads to the accumulation of sugars in the fruit. Thus, not only are sugar and acid content largely responsible for tomato sweetness and sourness, but the balance between them is an indication of ripening stage. The primary acids that make up the tomato are malic and citric acids, which can be estimated by measuring pH or titratable acidity. While pH typically fluctuates from about 4–4.5 in a tomato during ripening (Figure 3.1c), fluctuations in titratable acidity are much larger, ranging from 80 to 110 units of titratable acidity (Gautier et al. 2008). Both parameters are relevant to flavor, especially when measured in combination with sugar content.

Titratable acidity measurements typically require adding a known basic solution to the solution of interest (e.g., tomato juice) until the solution transitions from an acid to neutral. The amount of base added to neutralize the solution is then a measure of how much acid was present in the original solution. The titration process takes time and is best suited to lab or benchtop situations rather than in-field or other in-situ measurements. In contrast, pH is much easier to measure and a wide range of sensing technologies and instrument designs are available to measure pH in the field.

pH is useful for estimating flavor and also for monitoring the food safety level of processing tomatoes so that they can be harvested before they require that citric acid be added during processing to guarantee safety of the final product (Anthon et al. 2011).

OTHER TOMATO CHARACTERISTICS

There are other characteristics of tomatoes that play a role in the overall quality of tomato, although to a lesser degree than those characteristics already discussed. The shape of the tomato is important and, like color, influences the consumer's perception of the goodness of the fruit. However, tomato shape is largely determined by breeding and genetics and varies little within a particular crop (Bertin and Génard 2018). Thus, sensing shape is not particularly worthwhile within a crop.

Total soluble solid (TSS) content is also used to characterize tomatoes throughout maturity and ripening. TSS represents the percentage of the fruit that is made up of solids. Since a tomato is mostly water, TSS in a typical tomato varies relatively little, between 5% and 10% (Barrett et al. 2010), and can influence the firmness and texture of the tomato as well as the flavor. Since most of the TSS in a tomato consists of sugars, however, measuring sugar content directly is a more accurate approach to capturing the flavor quality represented by TSS.

Other chemical compounds in tomatoes, aside from lycopene and β-carotene, are also important to the nutritional quality of the tomato. For example, polyphenols are phytochemicals that are also found in tomatoes and offer a range of positive health effects including protecting the heart and liver as well as exhibiting antiviral and antioxidant properties, which support good overall health. The most common polyphenols in tomatoes are chlorogenic acids and flavonoids (dominated by naringenin and quercetin) and are primarily found in the skin of the tomato. Tomatoes also contain ascorbic acid (Vitamin C), which further contributes to the nutritional quality of the fruit (Bertin and Génard 2018).

ENVIRONMENTAL PARAMETERS

Many environmental parameters also influence the quality of tomato crops.

The intensity and duration of light the tomato plant receives as well as the temperature and humidity to which it is exposed in the air are major players in determining the mass, quality, and composition of a ripe tomato. High but not extreme temperatures increase sugar-to-acid ratios and improve tomato flavor but also reduce the firmness of the mature fruit. Temperatures lower than 10°C are known to decrease the Vitamin C, phenolic compounds, and carotenoids in tomatoes, while very high temperatures (above 26°C during the growing season) decrease lycopene and β-carotene (Gautier et al. 2008). Vapor pressure deficit, a combination of temperature and humidity, tends to increase the likelihood of blossom-end rot while decreasing the likelihood of tomato cracking during ripening (Bertin et al. 2000). And, high sunlight intensity improves ascorbic acid levels as well as β-carotene, lycopene, and phenolic compounds. In fact, tomatoes harvested over a summer growing season can deliver up to 50% more Vitamin C than those harvested in the spring (Massot et al. 2010). Among these three primary environmental parameters, light is considered the most critical parameter in affecting tomato ripening and quality.

Not surprisingly, crop management practices also affect tomato quality. Water stress, salinity, and fruit pruning are especially influential. Pruning fruit from vines increases the size of the tomatoes left on the vine (Guichard et al. 2005) and also increases the ratio of sugars to acids in the harvested fruit (Bertin et al. 2000), thereby producing larger and more flavorful tomatoes. Tomatoes also thrive under irrigation strategies that involve water stress and salinity. While water stress and salinity decrease the size of harvested tomatoes and have a mild but detrimental impact on overall yield, they also improve the ratio of sugar to acids as well as overall sugar content, which results in a more flavorful product (Ripoll et al. 2014, 2016). High salinity also improves the nutritional value of tomatoes via increased Vitamin C, lycopene, and β-carotene levels (Frary et al. 2010), although no generalizable results regarding a corresponding increase in phenolic compounds have been reported. Drought and associated water stress improve phenol content, but the effects on carotenoid content are more mixed (Ripoll et al. 2014).

SUMMARY

Although an exhaustive list of all things that contribute to tomato quality is not possible within the scope of this chapter, it is clear that what makes a perfect tomato involves not only things seen (color, texture), touched (firmness), smelled (aroma), and tasted (flavor, mealiness, juiciness), but many things unseen (carotenoids including lycopene, Vitamin C, polyphenols, sugars, pH). Deciding which of these parameters to measure depends on how the consumer perceives quality as well as which characteristics of the tomato contribute to that perception. Selecting which sensors or instruments will be used to make those measurements considers many things, including cost, ease of use, and return on investment. Identifying where and how the Internet of Things can support sensors in collecting, interpreting, and sharing data involves an even larger number of factors, many of which are explored in the chapters that follow.

CHAPTER 4

Sensors in the Internet of Things

IT'S SATURDAY MORNING AT 7:30 and my phone is raising a ruckus. Silly me for thinking I could sleep in for a day. Farming doesn't work that way, especially during harvest, and when it comes to the ever-finicky tomato, which changes its mind from day to day about whether it is destined to become the centerpiece of the salad or a tasteless terror headed for the catsup bottle. For a small farm like mine, the catsup bottle or any other processed tomato product is not an option. In order to keep my financial head above water, it's going to take an impressive yield of bright red, perfectly shaped, tasty tomatoes headed to market in a timely manner to keep the farm afloat.

I roll over and take a bleary look at the phone. It has kindly stopped its noisy squawking, but the notification that prompted said squawking remains front and center on the bright screen. In its dispassionate way, the phone is eager to inform me, via my latest Tomato App, that we have reached day 1300 since I first planted my flocks of seedlings in raised beds on the open fields. Those 1300 days aren't 24-hour periods but rather growth equivalent days that include a combination of 24-hour temperature cycles, sunshine, and other factors that all contribute to the process of turning my little green friends red. In concert, dozens of robust temperature, moisture, and light sensors installed on the fields have been tracking temperature and daylight in the fields since the seedlings went into the ground. Among a dozen other functions, the app updates me daily as to how close to the ideal harvest point of 1320 equivalent days we are for the varieties I've chosen for this season. We'll start our harvest cycle any day now, but unlike many other crops on adjacent farms, our tomatoes wouldn't think of ripening all at the same time. Rather, they will require at least four rounds of harvesting and likely more as we vary the ripening stage according to their final destination. This could be the produce stand we keep open through the summer at the entrance to our farm, the local farmer's market, the more distant produce stands on the other side of the mountains, or a select few mom-and-pop grocery stores that will choose our produce over that of the larger farms in the area. It's enough to make my head spin.

Fortunately, my smartphone not only lacks a head, but it also never seems to be overwhelmed. It just calmly lays out how much I'll have to harvest and when and at what ripening stage to do it, depending on the orders and estimated demand we have for the varieties we grow. Back when I first started with the app, it would often put me on the hook for far more than my fields could deliver, but since we installed the cameras on the fields, the app now estimates our yield pretty accurately, including separating the imperfects from the perfects among our fruit so I don't get fooled by raw numbers.

Since it is now so close to the harvest, automatic notices are sent out to our pickers as well, hopefully not jolting them out of bed this morning as has just happened to me. Barring any major and unexpected weather event (like a hurricane rolling through the Pacific Northwest in the middle of the dry season), my workers will be here on Monday. They will be ready to gently pick each ready tomato from the vine for fresh packing while simultaneously stripping each vine of small or immature fruit whose only purpose in life appears to be to steal nutrition and life from the larger more productive fruit. This is a huge difference from the days before the magic Tomato App. In those days, we walked the fields, guessing by eye and feel when the ideal time for harvest was coming, and often getting it wrong by several days, causing problems for the workers as well as my own financial bottom line.

In addition to politely updating the time to harvest, my phone has now started squawking about something much more urgent. The temperatures on the field stayed low enough last night for just long enough to present a potential chilling threat. This is where my ever-finicky tomatoes get sensitive about temperatures in the 40s (°F). While they may continue to ripen to that ripe red color that lands them on the table of many a consumer, they also hide internal damage and a drop-off in taste that's unlikely to bring return customers my way in the next growing season. Magic Tomato App is telling me that after comparing the last week of temperature data with the latest research data for my region, the risk of chilling damage has increased from only 10% to over 50%. That is nothing to be trifled with, so my leisure plans for the day are instantly canceled. I don't have to update my calendar. Magic Tomato App automatically clears my schedule when my little red friends are in distress. In fact, I wouldn't be surprised if my friends and family hear from the app more often than they hear from me personally.

After I've silenced the squawking and assured my phone that I will take action, I roll out of bed, into whatever clothing matches the morning's cooler temperatures and head out to the fields to check out the damage. Tomatoes are sly and fragile creatures. They may look happy, healthy, and headed to ripe red, but their insides can be telling an entirely different story if they have sat through too much cooling in the night. I gently apply pressure to an appreciable sample of tomatoes, trying to gauge internal damage by the tomato's firmness without causing damage myself. I thought about getting some of those smart gloves that automatically measure firmness while stopping short from causing damage, but those gloves cost a little bit more than I'm willing to spend right now. I am well aware that there is only so much objective touching of tomatoes that one human being can do in a day. Unlike me, the sensor-equipped smart gloves would remain unperturbed by fatigue as they take measurements on the 100th tomato as if it were the very first. While all of the manual firmness data I'm collecting this morning goes only to my head and sometimes faulty memory, all the data from the smart glove systems heads automatically to the cloud. There, Magic Tomato App could crunch the firmness and solids content data alongside the temperature profile data to revise the chilling risk and adjust optimal harvesting time and scheduling accordingly. But, I'm just not an early adopter. I've been waiting for the more tech-savvy farmers to pick up the technology and see how it works out for them before spending the money on more of that whacky technology myself. For now, it's just me roaming the fields trying to assess the impact of an unusual string of cooler temperatures over the past week.

As I walk, I remember when I first started with Magic Tomato App. At the time, I thought it would be a total waste of money. The temperature, light, and moisture sensors alone had set me back quite a bit, although not as much as I originally thought. I griped, moaned, and mocked the app for the first season. But, despite my ridicule and disdain, my sensors continued to chug along, interacting far more frequently with the cloud and accessing more current research and weather data than I ever could. At the end of the first season, my yields had increased by more than 20%. My mocking ceased. The next season, I installed a few cameras in the fields, and their yield predictions, along with early blight, insect, and disease alerts, made a world of difference not only in my yields but in the ever-present tomato anxiety that hangs over the head of every fresh tomato farmer who has ever economically lived to tell about it.

And so the morning goes. When I return from the fields, my phone has ratcheted the chilling risk down to 15%, so I won't have to change any of our harvest planning. I've collected a few samples that I'll test with my old-fashioned penetrometers just to make sure, but it looks like the crop is still in good shape for a bumper harvest.

How did I ever survive without Magic Tomato App? Thank goodness those engineers got a clue and figured out how to get all the right sensors onto my fields, integrate all that data into a package that my phone can tell me about anywhere, and do it at a cost that understands both our finicky crop and tight profit margins. I remember when those engineers first showed up, high on the Internet of Things (IOT) and selling us as many sensors as possible just because they were there. After a while and a ton of failures that spoke as loudly as the absence of success stories, the engineers seem to have realized how sensitive and particular fresh tomato crops can be and how much money we don't have to throw around. Of course, it helps that their technologies were first adopted by those larger, highly mechanized farms down south in California, where the tomato grown for fresh consumption roams the open fields about as rarely as the buffalo.

THE INTERNET OF THINGS UNVEILED

The Internet of Things is not just about adding more and more devices to the world just because it's possible. Instead, the IOT, when properly engineered, uses those devices to solve specific problems or to improve existing solutions to those problems in a way that such interconnectedness adds substantial value at minimal cost. The IOT is a massive, rapidly expanding market for electronic devices and information technology (Figure 4.1). Some now predict that capitalizing on the many interconnected devices that enable the IOT will soon surpass or even supplant technology development that has been driven for decades by rapidly advancing transistor and integrated circuit density dictated by Moore's law (Schaller 1997).

In agriculture, cost is often a major barrier to adopting any new technology, especially for small- to medium-sized farms. Thus, smart and cost-conscious solutions are the only solutions that are likely to enable the IOT to expand in agriculture and thrive in the world of tomatoes.

The IOT is about adding devices that can not only connect to each other and to the cloud but do so smartly. Its purpose is such that information can be collected and processed more quickly and efficiently and subsequent action can be taken to improve

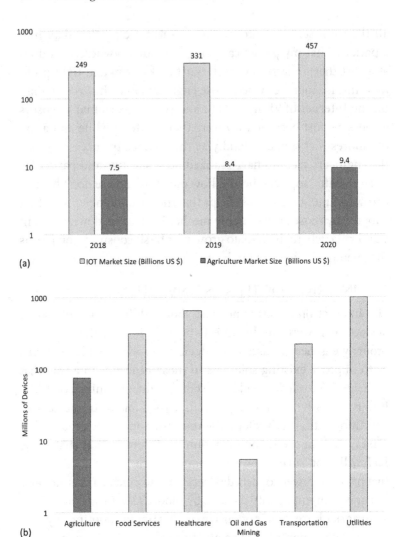

FIGURE 4.1 The rapidly expanding Internet of Things. The number of devices connected to the infrastructure that is the Internet and the profit reaped by the sale of those devices will continue to grow rapidly. Unfortunately, the application of IOT solutions to address challenges in agriculture lags behind other major sectors.

convenience, quality of life, and profit. The IOT should paint a picture of the physical world that is not possible with traditional devices that either connect only to the cloud or don't connect at all. A true IOT solution to a problem collects, transfers, analyzes, and acts on data that requires no human-to-computer or human-to-human interaction. Many IOT solutions also provide emergent capability that uses sensors, connectedness, and system design to solve problems in ways that individual sensors or devices that make up the IOT cannot.

For example, consider the IOT's entry into home automation. Wireless thermostats allow consumers to adjust temperatures in advance of walking through the door. Smart lights automatically turn on and off or dim in response to a person walking into or out of a room, as well as adjusting evening timing automatically to changing sunset and sunrise times. The ever-annoying chirp of the battery-starved smoke alarm can now be controlled with a single click to transform the chirp into a more palatable notification on the smartphone that alerts the homeowner of the need to replace the battery. Similar IOT-friendly smoke alarms can automatically contact emergency services when triggered. Smart refrigerators are able to track the contents and expiration dates of food items, making grocery lists a thing of the past. Smart blinds can automatically close at habitual bedtimes. And the list goes on. The IOT provides seemingly endless possible upgrades in convenience and efficiency to everyday living. But, in addition to providing these baseline improvements, the interconnectedness of these devices with other devices provides emergent capability. For example, smart thermostats, lights, and appliances can communicate with each other and with residential electricity meters to provide utilities with sufficient data to advise consumers on energy savings in ways never before imagined. This data, when collected in aggregate among many homes in a region, can also be used to optimize how much power is produced, at what times, and by what means, to reduce waste, diminish impact on the environment, and optimize reliability and cost of services.

While often seen as the next capitalist-driven vehicle of consumer convenience, the IOT also has the capability to contribute to solving many critical local, regional, national, and global problems. The emergent capability provided by many smart devices in the home to support better and more efficient electricity and other utility service and usage is just one example of how these devices can support a better and more sustainable world. The proliferation of wearable healthcare monitors provides another example of emergent capability that addresses not only quality of life but critical shortfalls in the healthcare system caused by ballooning costs and inefficient practices. Smart devices are available to monitor sleep, exercise, gait, eye strain, posture, blood sugar, heart rate, blood oxygen, and many other parameters that together can be used to enhance quality of life and extend lifespan well beyond what is possible within the present healthcare system in developing and developed countries alike.

Properly configured and strategically designed, the IOT can also increase the efficiency of agriculture and play a key role in supporting, scaffolding, and ensuring a global food supply that meets the needs of a growing world population. But, as great as the IOT sounds for its potential to improve crop health, yields, transport and distribution strategies, profit margins, and overall food volume, it must be applied smartly. It must be applied to what different types of crops actually need and what individual farmers, experimental farms, or farm collectives can afford.

Three basic types of IOT solutions are in use today. The first is a personalized IOT system that allows a user to set preferences so that IOT systems respond according to those preferences. The second involves passive-context awareness where the IOT system monitors an environment and provides users with options for action based on environmental data, and the third is an active awareness approach where a system automatically responds to environmental data (such as a fire or other time-critical crisis) with a predetermined action. In the IOT of agriculture, passive-context awareness systems are likely to be most applicable, providing calls for action to farmers and growers based on data collected from the field or greenhouse but stopping short of automatically taking those actions.

There are a number of passive-context awareness systems that have been successfully implemented in agriculture, but these are only the tip of the iceberg in terms of what the IOT can contribute to improving farming and harvesting practices and results. IOT can be applied far and wide to support greater food production, less waste, and a more economically fruitful bottom line for many farms, ranging from small to large.

THE INTERNET OF THINGS IN AGRICULTURE

Despite employing 1.4% of the labor force (BLS 2017) and contributing to 5.5% of the gross domestic product (GDP) of the United States (USDA Economic Research Service 2018b), the digitization of agriculture ranks dead last among major industrial sectors (Manyika et al. 2015). In advanced countries around the world, the strategic digitization of agriculture, exemplified by the Dutch (Viviano 2017), is a key pathway to the 70% increase in food production required by 2050 to feed a projected world population of 9.1 billion (FAO 2017b).

Production of food in developing countries will have to double by 2050 to meet global demand (FAO 2009). Around the world, the strategic use of technology to support more efficient farming operations and interconnection of these operations across rural communities also has tremendous potential to transform rural economies in developing countries and provide a viable and attractive alternative to urban migration (FAO 2017a).

However, increasing the use of technology or introducing an IOT approach to managing crops must be done with a clear benefit to crop yield and profit margins. This is especially true for farms that grow fresh tomatoes. In the United States, most of these farms are family operated and are small to medium sized. Among these farms, over 60% of small family farms and over 40% of medium-sized farms operate in the critical zone, with operating profit margins of 10% and potential financial problems (USDA ERS n.d.). The situation is similar at a global level. Around the world, agriculture employs over 1 billion people, many at and below poverty thresholds (FAO 2017b).

Given the stringent conditions and profit margins under which many of the world's farms operate, even those that specialize in high-value crops like tomatoes, an investment in the IOT requires a quick and significant return on investment. Adoption of technology by farmers, particularly the IOT, lags other industrial sectors, because farmers tend to have fewer resources to invest in technology and any funds used for such technology must be carefully and strategically invested. Success stories for the IOT in agriculture have ranged from strategic investments on the part of individual farms to far-reaching investments in databases and models that can be accessed at a national or even international level. While the IOT is often thought of as a solution that only applies to large farms and large sales volumes, it can also be relevant to boutique farms that use more specialized and less mechanized means to harvest crops. Examples of successful IOT implementations at these two ends of the farming spectrum follow.

Green Eggs and Ham (John Deere)

John Deere introduced the self-scouring steel plow to American agriculture in 1837. The plow allowed the American Midwest to be transformed into fertile farmland and enabled the John Deere company to grow into the largest manufacturer of plows in the world. Despite enabling this historic leap in agricultural productivity, John Deere lagged well behind other companies in marketing products and technologies for precision agriculture in the 1990s. Rockwell International and several other companies had already introduced global positioning system (GPS) technologies for mapping field location to crop yields and exited the business of doing so by the time John Deere began ramping up on similar technologies (Marsh 2018).

Sensitive to the bugs, inaccuracies, inadequate technical support, and expense that had dogged its competitors in marketing products for precision agriculture, John Deere first set about defining a killer app to guide its research and development of interconnected, digitization technologies. Not surprisingly, the company honed in on its original killer app—transformative improvements in plowing (Marsh 2018).

What resulted was Green Eggs and Ham, a yellow-and-green GPS receiver that sat atop its signature green tractors and collected location data from multiple GPS satellites. It corrected those signals using data from regional base stations using a C-band antenna and communicated with the mapping displays inside the tractor cab, all while withstanding temperature variations between −20° and 45°C, tractor vibration, and unwelcome visits from rodents and birds. Without GPS, tractors typically overlapped 1 meter of plowing between rows. With GPS, the overlap was much less, leading to significant drops in fertilizer used. In fact, by the early 2000s, John Deere's AutoTrac product was able to control a tractor so that it traced a line that was almost perfectly parallel to the tractor's previous path, reducing row overlap to almost nothing. This system provided not only a level of autonomy to plowing that had never before been feasible, but also allowed farmers to quickly and easily identify areas of concern in their fields. The result of this research and development effort, despite the late start, is that over 60% of North American farmers now use self-guided tractor systems, along with between 30% and 50% of European farmers and over 90% of Australian farmers (Marsh 2018). While accurate, effective, and broadly adopted, autonomous tractor systems are affordable only for large mechanized farms. For smaller farms, the IOT must take a different approach to be relevant.

Tackling Tasteless Tomatoes (Analog Devices)

A stunning, ripe, robust, red tomato may still be headed for the catsup bottle if it lacks the flavor that consumers now expect from fresh tomatoes. This is the problem that initially drew Analog Devices (ADI) into investing in New England tomato crops. The company's IOT journey with tomatoes began in New England with tracking growing degree days, which combine both temperature patterns and daylight to predict when a tomato is indeed ready to harvest for maximum flavor to go alongside its fine red looks. In its first use of the IOT in the open tomato fields, ADI installed temperature sensors with Bluetooth connections to local smart devices. Bluetooth eliminated the need for either readily available wireless internet or onboard power sufficient to transfer information over conventional wi-fi connections. Robust temperature-sensing packages installed in the fields collected detailed temperature data over time to accurately identify the 1300 growing day target. Collecting data, interacting with the cloud, and providing an app to track the growing day index provides farmers with a true IOT tool that gauges harvest time far better than traditional methods, which often rely on impressions more than on science. Detailed temperature data can also be used to predict and track the life cycles of insects and pests, and, in so doing, allow farmers to better manage those pests in pursuit of the perfect tomato.

In addition to IOT-scale temperature sensing, ADI also invested in widespread measurements of moisture, both above and below ground. Too much moisture above ground (watering, dew, humidity) can cause a wide range of problems. Tomato plants that are too wet too often are vulnerable to white blight mold and bug infestations, which can destroy entire crops. Tomato plants that are too dry for too long, on the other hand, can fail to thrive, thus leading to reduced yields and lower-volume fruit. Sensing moisture below ground complements these above-ground measurements by providing data necessary to optimize irrigation, minimize root stress, avoid vapor pressure deficit, reduce water usage, and limit chemical runoff to surrounding waterways. Like temperature sensors, moisture and humidity sensors are small and can be integrated into miniaturized, standalone packages that can communicate data to the cloud and provide analytics valuable to not only the farm and crop in which the data was taken but also in surrounding agricultural operations. However, while temperature has little impact on the performance of a temperature sensor package, moisture, which is all too common in the dew-filled mornings of the Northeastern United States, can confound moisture sensor readings and even damage or reduce the lifetime of the sensors themselves.

Temperature and moisture sensors are, in one sense, low-hanging tomatoes (fruit), in that they are well suited to agricultural problems, are inexpensive, and can be implemented on an IOT scale in a fairly straightforward way. To truly advance from a tasteless to a tasty tomato, however, requires monitoring the chemistry of tomatoes. Noninvasive, nondestructive sensors to monitor pH, fructose, salinity, glucose, sucrose, lycopene, and other chemicals and nutrients are as essential to optimizing taste as monitoring growth index and water imbalance are to optimizing yield. Together, taste and yield maximize economic benefit. ADI has demonstrated strides forward in the use of sensors in an IOT context for improving tomato yields via strategic use of dense temperature and moisture data. The ADI project has clearly illuminated the potential of the IOT to support better tomato farming. How far and in what directions the IOT can successfully go beyond temperature and moisture, however, remains to be seen (Insights 2016).

Summary

IOT success stories in the world of agriculture are a result of far more than interconnecting sensors in such a way that they provide useful data to the cloud. Rather, they are solutions that are based on the everyday realities of farming as well as the stringent cost, time, and other resource constraints that uniquely hamper farmers from adopting technology, no matter how valuable those technologies may be in the long term.

INTERNET OF THINGS TOMATO STYLE

Sensors are not new to the world of tomatoes, whether pre- or postharvest, whether fresh or processed. In order for the Internet of Things to gain a to-mato-hold in the fields, however, the IOT must offer quantifiable benefits from increasing both the number and connectedness of sensors that monitor those fields. While many types of sensors can provide new and timely insight into the maturity and health of a tomato crop, installing those sensors, particularly in small- to medium-sized fresh tomato operations, does not necessarily make economic sense. While profit margins for high-value crops like tomatoes are generally higher than those for mainstream crops such as corn, wheat, and soybean, they are still more restrictive than many other industries that seek to capitalize on the potential economic benefits of the IOT.

To be true to an IOT view of engineering solutions in agriculture, sensors must address a number of criteria (Table 4.1). First, they must demonstrate *performance* suitable to real problems in real agriculture. These sensors should also be *compatible* both with the autonomy expected of an IOT-scale solution and with existing agricultural platforms (stationary, wearable mobile) that can host them. Furthermore, sensors that underlie IOT solutions should also be generously distributed in the fields (or greenhouse) or, alternatively, take frequent measurements over time in order to provide *dense* data that traditional sensing and monitoring techniques cannot provide. Beyond these foundational characteristics of IOT solutions, sensors developed for farming operations must also be *accessible* in terms of commercial

TABLE 4.1   What Should a Sensor Deliver to the IOT?

	Description
Accessibility	Commercial availability, affordability, battery life (power consumption), size (compactness), and ease of use (including user interface, interconnectedness, and maintenance)
Compatibility	Invisible, seamless, wireless, autonomous connection to the cloud
	Compatible with existing stationary and mobile platforms used in agriculture
Density	Capable of providing large amounts of data
Performance	Accuracy, durability, dynamic range, resolution, stability
Versatility	Easily adaptable to other problems, multifunctional

availability, cost, power, and other constraints and be *versatile* so that the data they collect can be adjusted to sense and address multiple problems in farming operations. Without all of these qualities, IOT solutions are not likely to be adopted, even by the most tech-savvy farmers.

Accessibility

In order to be accessible to IOT-scale farming (whether to individual farms, experimental farms, or other organizations or collectives that collect data to support multiple farms), sensor technologies must be available, affordable, and manageable in terms of cost, power consumption, size, and weight. Many sensor technologies, despite bearing great promise for a wide range of IOT applications, are still at the research stage or are available only as limited prototypes. Thus, their commercial availability and widespread accessibility to growers are limited. In addition to commercial availability, affordability also limits the risks that tomato farmers can and will take in adopting new technologies. Small to medium farming operations, in particular, are the norm among those tomatoes harvested for fresh consumption, and many do not make a profit or are supplemented by alternative forms of income (USDA ERS 2018a, USDA ERS n.d.). A sensor network managed by an IOT solution must earn its keep among these smaller farms. Sensors that run into the hundreds of dollars or fail to leverage existing technologies (e.g., smartphones or established short-range communication protocols) are usually out of reach. Even among mechanized farms that harvest larger crops for processed tomato products, cost remains an issue, as profit margins remain tight. If cost and availability concerns are addressed, sensor technologies must also meet expectations for low power consumption and be of reasonable size and weight to be used on a range of platforms in the fields, in greenhouses, in storage, postharvest handling, or transport. And, finally, accessibility also embraces ease of use. No matter how amazing, any technology that requires too much maintenance or is too difficult to use will find itself quickly collecting dust in a remote corner, forgotten and abandoned.

Compatibility

Sensors in an IOT-based approach to improving the pre- and postharvest processes associated with growing tomatoes must be compatible both with the basic characteristics of the IOT and also with existing platforms on farms and growing operations that can potentially host sensors.

In general, the IOT relies on sensors operating autonomously with little human intervention. But, even on mechanized farms, autonomy is not a word frequently associated with the sensors and instrumentation used to monitor the health and well-being of the growing tomato. But, within the IOT, sensors are expected to operate on their own with little human-to-human or human-to-computer support. In an ideal world, the IOT runs the whole show among the tomato vines. It monitors the health of the crop from the moment the seedlings reach the ground, staying silent until crop health is at risk, irrigation has gone awry, or harvest is imminent, except for a daily report to the growers and farmers to assure all is well or to alert a cause for action. Sensors can be autonomous in many different ways. They can be deployed and operate far removed from a human being altogether. They can silently piggyback on other human activity in the fields, whether integrated onto tractors or seamlessly woven into gloves. Autonomous sensors can hitchhike on drone flights or be deployed on futuristic unmanned ground robots that roam the fields and protect every growing tomato from anything that might interfere with perfection at harvest time. But, most importantly, autonomy is an essential ingredient to sensors in the IOT because it enables vastly more data to be collected than what is possible when human intervention is required. Autonomy supports greater and more cost-effective data collection, as there are no labor costs involved in collecting such data. Ideally, an autonomous sensor network collects data with little maintenance; is almost invisible to the farmer or grower; and provides only the processed, aggregated data to the end user in an easy-to-use, magical smartphone app or similar friendly user interface. A major challenge of building practical and adoptable IOT systems for sensing the perfect tomato is often not in finding the right sensor to collect the right data. Instead, the IOT challenge is in building the smart system to autonomously aggregate, sort, interpret, and validate results quickly enough to support strategic action for improving what is ultimately harvested and sold. Sensors that support an IOT system must also fit within and be compatible with its footprint. An imager that doesn't fit behind the lens of a camera is not useful. A remote temperature sensor that has a 2-hour battery life is not practical. Tomato chemistry on a chip that can't communicate with other devices or with the cloud is destined to stand alone. Certain sensing platforms fit the IOT and certain platforms don't.

In addition to being compatible with the IOT, sensors used in precision farming must also be compatible with what is available on and near the fields (or greenhouses). In the world of growing tomatoes, options for sensing platforms are limited. Internet access, whether wired or wireless, may be intermittent or of limited bandwidth. If a sensor sits stationary on a pole or is otherwise integrated into the landscape, it must be able to collect data and remain flexible as to when and how to transfer it to the cloud. Some stationary sensors will have the luxury of transmitting in real time across an efficient wi-fi connection. Others may have to transmit data to another sensor, hopping across nodes using local wireless connections like Bluetooth before the data reaches its final destination for analysis and action. Other sensors may have to store data and opportunistically sense time and mode of transfer, whether via a person walking by with a smartphone, a drone doing a fly-by, or something in between. Stationary, in-situ sensors also have the added burden of being able to withstand a range of weather conditions, day after day. If a sensor is on board a drone, it must be able to know its location and strategically collect data to conserve power. A sensor on board a tractor is tasked with other challenges. And, the big and bulky techniques that characterize many traditional modes of sensing crop health, optimal harvest times, and other relevant parameters to the healthy tomato are inherently facing an uphill battle to be compatible to an IOT approach to farming. Many of these big and bulky techniques are reserved for benchtop, off-site locations, and their efficient, cost-effective use relies on sharing data with other farms and stakeholders, including research facilities and experimental farms.

Density

One of the main attractions of sensors installed within an IOT context for monitoring crops is the density of information that can be collected and interpreted by these sensor networks. Feeding information back to individual growers about individual fruit on a daily basis may seem futuristic or even out of reach. But, in the high-value business of growing fresh tomatoes for consumer consumption, every tomato is worth saving, and finding cost-effective sensors to facilitate higher yields is often worth pursuing. Traditional crop monitoring techniques have tended to prefer off-site testing of select fruit to infer conclusions about an entire crop. These traditional methods can be complemented or replaced by sensor networks that are designed to collect equally accurate but much denser information about tomatoes without overloading the information processing capacity of individual farms. Whether at a local individual farm, a regional, or even a national scale, dense data is a hallmark of IOT solutions, and in supporting and searching for each perfect tomato roaming the farms of the United States, collecting data at both high spatial and temporal resolutions is a vital part of the process.

Performance

Any sensor technology used in a viable IOT solution must perform to the requirements of the application or problem to be solved.

Performance specifications relevant to sensing the perfect tomato include but are not limited to:

- *Accuracy*: Does the sensor provide a measure of the target environmental parameter within desired limits?

- *Durability*: Can the sensor withstand typical use (e.g., droppage) and exposure (e.g., temperature swings, humidity, weather)?

- *Dynamic range*: Can the sensor measure, from minimum to maximum, the desired environmental parameter with the desired accuracy?

- *Resolution*: Does the sensor precision allow for measuring the targeted environmental parameter within the desired increments or decrements?

- *Stability*: Does the sensor output remain stable over key measurement periods, or does it drift too much?

These parameters are relevant to many but not all sensor needs pre- and postharvest. Other performance metrics, such as hysteresis and selectivity, may also come into play for particular sensor problems.

Versatility

Although the chapters that follow focus on sensing specific attributes of the perfect tomato, any sensing technology deployed to this task should also be versatile enough to perform other functions. For example, cameras or imagers used to determine the color of an individual tomato for optimal harvest should also be able to assess other qualities of crop health, including weed density, blight, irrigation abnormalities, compromised quality from chilling, insect damage, and other threats. pH sensors used to support food safety while tomatoes are awaiting harvesting and processing must also be useful to assess soil pH and also be used in conjunction with other parameters to predict taste and nutritional quality. To be cost effective and earn their keep among the tomato vines, sensor networks working within the IOT must be versatile, able to reconfigure and adapt to multiple problems in the fields that go well beyond optimizing harvest time during desired phases of the maturation and ripening process. To be true to the promise of the IOT, data collected by sensors must also benefit from being highly interconnected. For example, data from one growing season can be used to inform decisions during the next season at the individual farm level. Alternatively, such data can support research of regional experimental farms and applied labs that seek to optimize both yields and crop choices across an entire region or microclimate.

In summary, versatility both in sensing capability and in data interconnectedness can enable quick adaptation to emerging challenges in farming not only on a single farm or in a single season, but across entire regions of crops across many seasons of farming and across a range of crop choices. Ultimately, the versatility of sensors for the inclusion of the IOT in agriculture, including flexibility and reconfigurability, are essential, at a global scale, to stimulate better and more efficient farming practices at the pace required to keep up with global food demand.

SENSING FOR THE PERFECT TOMATO

There are as many different opportunities for and approaches to sensing for the perfect tomato as there are varieties and cultivars of tomatoes consumed around the world, from fresh and whole to canned and diced to catsup and juice. Some sensors, like those that sense temperature, are inexpensive and robust enough to be implemented on individual farms in large numbers and have the potential to collect, store, and transfer high-resolution data over space and time into the cloud for determining equivalent growing days and predicting time to harvest. Some sensing technologies are equally well suited to individual farms, like on-field cameras, but have yet to be both engineered to meet the challenging conditions of open fields and tailored to the unique needs of imaging particular crops from planting to harvest. Still other sensing technologies, like hyperspectral imaging, are better suited to experimental farms or select crisis scenarios, due to high capital costs, complexity of use, or other adoption barriers.

Addressing all potentially useful or appropriate sensing technologies for monitoring of tomato quality both pre- and postharvesting is beyond the scope of this book. Some sensors, such as those that capture temperature and moisture, have already proven successful at an IOT scale in tomato farming (Insights 2016). Other sensors, while conceptually useful to tomato farming, have not yet arrived on the farm in an IOT context. The following chapters focus on sensing three parameters (color, firmness, and pH) that have the potential to transform tomato farming if implemented within the IOT and in a cost-effective manner. This sampler platter of sensors invites the reader, whether on the farm, in the classroom, or somewhere in between, to consider the tantalizing possibilities for the IOT in tomato farming.

Chapter 5 begins this exploration by looking at the possibilities for monitoring individual tomato color from the moment a tomato reaches mature green to the point it reaches the consumer's hands. Monitoring color also has broad applications outside of the tomato itself to monitoring leaf health, soil variability, and other parameters in both open fields and greenhouses. Thus, versatile and accurate color monitoring can be extended well beyond the tomato to other soil and plant health challenges. While monitored far less frequently, firmness (Chapter 6) is also an important up-and-coming parameter for the IOT and can be used not only in preharvest decisions and evaluation but also in postharvest handling applications. Many sensors that can measure firmness can also gather additional data to infer other internal properties of the tomato, including soluble solids content and emerging bruising. And, finally, to go out on an acidic limb, pH sensing (Chapter 7) completes the sampler with its importance to tomatoes grown for both processed products (for supply chain optimization) and fresh consumption (for optimizing taste).

Considering the technologies that can enable sensing of color, firmness, and pH can at best put a meaningful scratch on the surface of what is possible with the IOT in tomato farming. How these sensor technologies are implemented in farming operations on both experimental and individual farms in the future, however, will depend on how well IOT application engineers can understand and engineer these technologies to meet the needs and honor the constraints of the perfect tomato.

CHAPTER 5

Sensing for the Color of the Perfect Tomato

COLOR CHANGES DRAMATICALLY AS most tomatoes mature and ripen. These color changes have been segregated and standardized into six discrete stages (USDA 2005) for traditional red tomato varieties:

- *Green* is the initial ripening phase of a typical tomato, where the surface of the tomato is entirely green, although the shade of green can vary from light to dark.

- *Breakers* include tomatoes that show shades of yellow, pink, or red on up to 10% of the tomato's surface.

- *Turning* is the ripening phase where between 10% and 30% of the tomato's surface is yellow, pink, or red, and the rest of the tomato remains green.

- *Pink* refers to the fourth ripening phase where between 30% and 90% of the tomato is no longer green and is showing shades of pink or red.

- *Light Red* means that pink or red covers over 60% of the tomato surface but red covers no more than 90%.

- *Red* refers to a ripe tomato where over 90% of the surface of the tomato is red.

These ripening stages can be adjusted for tomatoes whose end-stage colors are not red but rather orange, purple, black, yellow, mixed, or striped and also progress from mature green through several phases before reaching a fully ripe color (USDA 2005).

Color has long been an important characteristic of the tomato. Tomatoes are harvested at different colors and maturation stages depending on the end use of the tomato, whether for fresh consumption or processed products (Monti 1979). In some cultures, growers make harvesting decisions based entirely on tomato color and biological age as measured by the time after anthesis or flowering (Tran et al. 2017). While a key contributor to determining time to harvest, color is also used to assess tomato quality. For example, consumers use color and appearance to choose the fresh tomatoes they purchase even though the external appearance of a whole tomato is not always an accurate indicator of its ripeness or flavor (Shewfelt 2000). Furthermore, despite the increasing availability of alternative tomato varieties, consumers still expect a tomato to be uniformly red, not orange, purple, or multicolored (Barrett et al. 2010). The redness of a tomato is also strongly correlated with lycopene content (Arias et al. 2000, Davis et al. 2003), one of the tomato's most distinctive and valuable nutritional benefits. In addition to expressing lycopene, tomato color also indicates the balance of other pigments present in the fruit, including fat-soluble chlorophylls; yellow, orange, and red carotenoids in addition to lycopene; water-soluble anthocyanins (red); and flavonoids (yellow) (Barrett et al. 2010).

Assessing color, whether by machine or by human eye, is an important part of sensing for the perfect tomato. However, using technology rather than human perception to assess color enables objectively distinguishing maturation and ripening well beyond the six discrete stages identified by the USDA. Moving beyond these six discrete stages allows for tracking biological variations within a single plant and harvesting for maximum yield and minimal crop loss on a tomato-by-tomato basis. Since tomatoes ripen at different rates, even on the same plant, it is important to monitor individual tomatoes to determine optimal harvest times. The same is true for postharvest handling of tomatoes grown for fresh consumption because color is critical to quality during sorting, storage, and transport.

Traditional techniques for measuring color using spectrophotometers or colorimeters are one dimensional and often destructive, thus making them ill suited for widespread measurements of tomatoes on the vine. Many methods for sensing color are relatively inexpensive and can be cost effective not only for experimental farms, which are invested in modeling of different tomato varieties, but also for individual growers who are interested in improving specific, localized harvest practices. However, advances in imaging technologies have opened up new possibilities for nondestructive testing of tomato color. This chapter considers the most viable technologies (Figure 5.1) for doing so, ranging in complexity from single-point measurements on a single tomato to multidimensional imaging of an entire tomato crop.

TRADITIONAL APPROACHES TO COLOR MEASUREMENT

Spectrophotometers are the instrument of choice in high-resolution, precise, and objective measurements of color. The color of any object is determined by the types of light (i.e., wavelengths) that are reflected off the object. Red tomatoes reflect primarily red light, while mature green tomatoes reflect both red and green light, thus giving the appearance of being light green. A spectrophotometer typically measures reflected light from an object at a fixed angle, typically 45°, to the incident light. Variations in this technique include measuring the light reflected at all angles (by using a spherical instrument), or a multiangle approach that

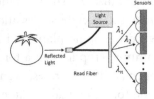

(a) Color Sensor   (b) RGB sensors on CMOS Imager   (c) Spectrophotometer

FIGURE 5.1 Approaches to detecting color. Color can be measured using (a) individual color sensors that measure broad wavelength bands (e.g., red, green, blue) at a single location (both low spatial and low wavelength resolution); (b) two-dimensional imagers of those color sensors (high spatial but low wavelength resolution); or (c) spectrophotometers (low spatial but high wavelength resolution).

measures color at different angles of incident light to understand how perception of color may change depending on how the object is viewed. Typically, the basic fixed-angle technique is sufficient for measuring the light reflected by a tomato. This reflected light is then split into its component wavelengths or colors and the resulting spectral footprint used to differentiate good-quality from poor-quality tomatoes, both before and after harvest (Figure 5.2a).

Spectrophotometers are available in both benchtop and portable/handheld models. Benchtop instruments tend to be more expensive but provide more detailed and precise color information than handheld instruments. Benchtop spectrophotometers are available from Hach, ThermoFisher Scientific, Mettler Toledo, and several other vendors and range in price from US$2000 to US$10,000. Hunterlab makes a spectrophotometer specifically for measuring the color of tomatoes and tomato products, the ColorFlex EZ Tomato Spectrophotometer. Benchtop instruments deliver color detection resolution on the order of nm (10^{-9} meters) for wavelengths in the visible spectrum and beyond, across dynamic ranges from UV (10–400 nm) to near-infrared light (780–2500 nm). In general, these instruments are not portable and require an AC (nonbattery) power source. Portable or handheld

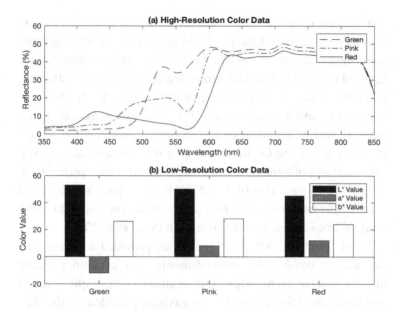

FIGURE 5.2 Representations of tomato color. Color can be determined by measuring reflected light (a) in a high-resolution wavelength spectrum or (b) in low-resolution, aggregated wavelength bands that approximate the sensitivity of the human eye.

spectrophotometers are also in plentiful supply, available from Hach, Ocean Optics, and others, and range in price from about US$1000 to over US$3000. A few of these devices are battery operated but differentiate color at lower-wavelength resolutions than benchtop instruments, or provide high accuracy but only at a small number of fixed wavelengths. Nevertheless, portable and benchtop spectrophotometers are the standard in objectively determining color with excellent precision and accuracy. Unfortunately, both benchtop and portable instruments are expensive and usually require damaging the tomato under test in the process of measuring its color.

A less expensive alternative to the spectrophotometer is a colorimeter, which measures a small number of wavelength bands (colors) consistent with the human perception of color. The CIELAB color space, introduced by Commision Internale de L'Eclairage, provides a unified and standardized framework for making these color measurements along three axes of perception: light to dark (L*), red to green (a*), and blue to yellow (b*). Measurements along these three axes vary during ripening (Figure 5.2b). The ratio between a* and b* (Arias et al. 2000), the hue angle calculated from a* and b* (Hobson et al. 1983), and all three axis values (López Camelo and Gómez 2004) have been used to differentiate various stages of tomato ripening and determine the optimal time to harvest. Colorimetry in the CIELAB color space has also been successfully used for monitoring tomato firmness (Batu 2004) and total soluble solids inside the fruit itself (Saad et al. 2016).

While the CIELAB color space has provided a framework for standardized color measurements that are independent of colorimeter technology and manufacturer, light sensors commercialized for consumer cameras have provided another key opportunity for detecting color. The proliferation of these sensors in consumer electronics has enabled very low-cost and highly portable color sensing. However, these sensors do not detect color along the same axes as CIELAB. Instead, these light sensors have three filters that isolate red, green, and blue (RGB) components of incoming light for purposes of duplicating the photographic functions of traditional analog cameras and mimicking the sensitivity of the human eye. In some cases, RGB has proven as effective as CIELAB colors in analyzing fruit ripeness. For example, the R value alone has been shown to be useful for identifying the major ripening stages of tomatoes as well as fruit firmness (Schouten et al. 2007), while R and B values in combination with fuzzy logic have been used to classify ripening stages of tomatoes with over 94% accuracy (Goel and Sehgal 2015). In some studies, both RGB and L*, a*, and b* values have also been measured to demonstrate comparable results between the two approaches to measuring color. These studies have shown that RGB measurements can produce comparable results to traditional L*a*b* colorimeter or chronometer values when measuring tomato ripeness or maturity and inferring Brix and lycopene content from those measurements (Niño-Medina et al. 2013, Takahashi et al. 2013).

Advances in integrated circuit and microfabrication techniques have enabled a new generation of color sensors that will continue to drive down the cost and improve the portability of colorimetry for studying color in the field. Some color sensors, such as the TCS34725 (manufactured by AMS AG), take a passive approach to color sensing, using infrared blocking filters, sensors with reduced sensitivity to ultraviolet light, and a clear light sensor to supplement RGB sensors in order to detect color more accurately in varying ambient light conditions. Other sensors, such as the SPECTRO-3 color sensor (sold by Sensor Partners) provide options for both passive and active color sensing. In active color sensing mode, a light source integrated with the sensor itself illuminates the surface of interest, and the reflected light is measured relative to this light, thus desensitizing the measurement to ambient light. Active sensing mode provides more consistent measurements, while passive sensing mode consumes less power.

In summary, relatively expensive spectrophotometers are likely to remain limited to small numbers of destructive measurements on tomatoes whether before or after harvest. Colorimeters and chronometers, aided by the decreasing cost, proliferation, and advancement of color sensors, are promising for frequent measurements of many tomatoes in a batch or crop. At present, commercially available colorimeters are primarily limited to handheld devices that can be manually operated to monitor tomato color and infer ripeness and other valuable parameters. However, in principle, there is nothing limiting these same color sensors from being deployed on more seamless platforms (such as smart gloves) or on mobile platforms to better support an approach to comprehensively measuring and sharing tomato color data.

CONVENTIONAL IMAGING TECHNOLOGIES

Spectrophotometers and colorimeters use mature technologies and provide an accurate means to measure tomato color and properties such as firmness, soluble solid content, and chlorophyll that are strongly correlated to color. In recent years, several highly accurate, small, and inexpensive color sensors have been commercialized. Advancements in color sensing technology are likely to continue to drive down the cost and increase the accessibility of sensing color in agricultural applications.

At best, a single measurement from a colorimeter or spectrophotometer evaluates a single fruit at a time. At worst, by measuring only a small portion of a single fruit, such a one-dimensional measurement provides an inaccurate assessment of that fruit's ripeness, readiness for harvest, and overall quality. Since harvest readiness varies widely even within a single tomato plant and especially for indeterminate varieties, one-dimensional, limited quantity, small area sampling used to assess optimal harvest time is often inadequate when used alone. To supplement or replace one-dimensional measurements, the next logical alternative to evaluating color among a crop of tomatoes is to use two dimensions, capturing an image rather than a single point of color. Postharvest, two-dimensional imaging is also valuable for better understanding batches of tomatoes for sorting, transport, and distribution.

Two-dimensional assessment of color can be provided by the same imagers and imaging technologies used in digital cameras. Two primary technologies are available for commercial imagers: charge-coupled devices (CCD) and complementary metal oxide field effect transistor (CMOS) devices. Both technologies rely on the photoelectric effect where incident light falling on the light sensor donates sufficient energy to electrons in the underlying semiconductor to change the sensor output. In this process, electron-hole pairs are generated that push the electrons up to the higher energy levels in the conduction band, thus increasing the freely available current carriers in the semiconductor sensor and increasing sensor current. Semiconductors used in both CCD and CMOS designs are sensitive only to light whose wavelength is shorter than a maximum wavelength, which is determined by the energy bandgap of the semiconductor. Common semiconductor materials such as silicon (Si) and gallium arsenide (GaAs) have energy bandgaps that allow sensitivity to both visible and infrared light.

Older imager technologies use CCDs which collect the electrons generated by the photoelectric effect in each pixel and pass the charge down rows and columns to an off-imager site where the charge is buffered and converted to an analog voltage. CCD signals are often transferred and converted in this way one pixel at a time, which limits their overall speed. In contrast, CMOS devices work by converting electrons generated by the photoelectric effect directly and immediately on the pixel itself, often into digital form, before passing these pixel values off chip. CCDs have the advantage that almost the entire pixel area is dedicated to collecting light, while CMOS devices use only part of the pixel for light collection while the rest is used for signal conversion. Since each pixel on a CMOS imager does its own conversion, variation in pixel-to-pixel output is greater in CMOS imagers than in CCD imagers. While CCDs offer improved uniformity, fill factor, detection limit, and sensitivity, CMOS imagers are faster, which can improve throughput in conveyer belt scenarios. CMOS imagers can also be fabricated using more standardized microfabrication techniques used for integrated circuits, thus making them less expensive than CCD imagers (Teledyne Dalsa n.d.). CMOS imagers are also naturally resistant to blooming (spilling of light from one pixel to neighboring pixels in bright conditions), thus making them a better choice than CCD arrays to avoid overexposure in bright sunshine or other highly illuminated environments. Combined with lower power consumption and ease of integration with other functions and circuitry, these benefits make CMOS attractive for low-cost consumer electronics applications, and these imagers have proliferated rapidly over the past decade. While CCD imagers are still preferred for specialized applications such as remote sensing and near-infrared sensitivity, CMOS imagers have become the technology of choice for mainstream imaging. Demand for low-cost and high-resolution imagers continues to increase, which will continue to drive costs down, making imagers accessible to more and more applications in and around agriculture, including the monitoring of fresh and processing tomatoes (Thusu 2012).

For the high-spatial-resolution, high-pixel-count images they provide, imagers are low cost and readily implemented either on stationary or mobile platforms situated on individual or experimental farms. However, opening the field up to two-dimensional imaging poses some challenges. First and foremost, most low-cost imagers are not designed to measure true color but rather to duplicate human perception of color by collecting RGB information and transforming it into a single image using the luminosity curve of the human eye as a guide. While this approach to two-dimensional measurement correlates inherently well with human perception of tomato color, it is not well suited to capturing true color. And, since the human eye is far more sensitive to green than to red, conventional cameras will tend to attenuate or de-emphasize the key color in a ripening tomato—red. However, closer observation of the reflectance spectrum of a tomato (Figure 5.2a) reveals that it is actually the loss of green wavelengths that characterizes the ripening process more than the addition of red. Thus, while imagers are limited in their ability to deliver true color information, they are fortuitously well designed to capture the significant changes in tomato color that characterize the ripening process. Furthermore, because the quality of an imager in a camera is directly related to its ability to duplicate what the human eye sees, a high-quality imager will deliver consistency in color duplication, limiting variability between cameras. Low batch-to-batch variability among imagers has emerged from multiple decades of evolving and competing product designs among digital camera manufacturers. Thus, within a given line of imagers or cameras, device-dependent variations in color representation and measurement can be quite small. The result is predictable and consistent performance that provides a golden opportunity for farmers to track the ripening process and postharvest changes in their own tomato crops and also to correlate this data with similar data shared to the cloud by other farms as well as model data developed by experimental farms and research efforts. Sharing, comparing, and leveraging data promises greater yield, more optimal harvesting, and overall improved economic results for individual farms as well as regional markets.

Two-dimensional color measurements obtained using digital cameras must also contend with the same variations in background lighting that often limit the quality of outdoor photography. Perception of color can change dramatically with variations in ambient light. The color of a tomato can be perceived very differently depending on cloud cover, time of day, dew cover, weather conditions, and angle of view. Some of these variations can be compensated for during the collection of an image. For example, auto-exposure changes aperture and shutter speed so that a predictable amount of light reaches the imager, providing consistency in the overall highlights, shadows, and middle tones of the resulting images. However, auto-exposure does not account for changes in the color of ambient outdoor lighting, which can vary with cloud cover, weather, and time of day. For example, sunlight through cloud cover contains less green, yellow, red, and infrared content than clear sky sunlight, regardless of whether the sunlight is measured directly or indirectly (Lee and Hernández-Andrés 2005). Infrared filters fitted onto imagers or color sensors can reduce these variations and limit the impact on color perception; however, complete immunity to ambient light is impossible, and attempts to mitigate this source of error may include measuring cloud cover and taking steps to capture images under similar lighting conditions from day to day. Alternatively, cloud cover data can be collected from sources in the other cloud (cyberspace), whether on-site or off, alongside other weather conditions and temperature and humidity information. This data can be used to adjust images to enable more accurate comparisons of day-to-day ripening changes. It can also be used to make more accurate comparisons between tomato crops grown on individual farms vs. those grown for modeling and characterization purposes on experimental farms.

While variations in intensity and color of ambient lighting limit the ability to track tomato ripening over time due to inaccuracy in RGB measurements, spatial variations in crop images can also limit the usefulness of imager technologies in tomato crop monitoring. Even in a stationary setup (e.g., camera mounted on a pole), variations in the spatial location of individual tomatoes in an image are inevitable due to plant growth, wind, and other disturbances. A host of image processing algorithms and techniques have been developed to compare images. These techniques make it possible to monitor color changes on a day-to-day basis for individual tomatoes. However, the complexity of image and color monitoring tasks increases dramatically when a mobile platform is involved. Whether mounted on a manned vehicle such as a tractor or an unmanned vehicle such as a drone, using images collected while moving makes it difficult to compare results from day to day at the individual tomato level. As with stationary platforms, imagers on mobile equipment are affected by natural disturbances that create small changes in the location of each tomato plant. Compounding these localized variations are inevitable changes in the vehicle's pathway through the fields and uncertainties in determining the exact location of the vehicle when an image is taken. Therefore, in the fields, mobile monitoring platforms are best suited to problems that require understanding population characteristics rather than individual tomato characteristics. For example, mobile cameras can be used to estimate the overall fraction of a crop that is ready for harvest or to monitor the extent of insect and other damage to both plants and fruit.

Postharvest, imaging problems are simpler because the environments in which imagers are used are better controlled than in open-field situations. Imagers can be conveniently mounted over conveyer belts to monitor the color of individual tomatoes. They can also be used during storage and transport to monitor postharvest treatments or to track degradation and over-ripening.

In these scenarios, imager data can be collected using mobile platforms with little loss of accuracy over stationary platforms.

ADVANCED IMAGING TECHNOLOGIES

A conventional imager captures three broad bands of color (red, green, and blue) in a two-dimensional image. Because red, blue, and green color sensors cover a broad and overlapping range of wavelengths (colors), the result is a color image, but one in which multiple combinations of colors (i.e., light intensities at different wavelengths) can add up to an image that looks the same to the human eye despite containing underlying differences in color. Thus, conventional cameras produce inherently ambiguous images. This ambiguity matters little when a photograph is only intended to duplicate what the human eye sees. However, when the imager is applied to problems where knowing true color is critical, this ambiguity becomes more of a concern.

Hyperspectral imaging produces a three-dimensional output that largely resolves this ambiguity. At every location (or pixel) in an image, a detailed spectrum of color (wavelength) at that image is extracted. For example, in the visible wavelengths that range from approximately 390–700 nm, sampled in 1-nm increments, a hyperspectral image of resolution $M \times N$ (pixels) produces a hyperspectral cube consisting of 310 images, each of $M \times N$ spatial resolution.

But, hyperspectral imaging (HSI) is both expensive and power hungry. Taking many images at different wavelengths also takes time, which requires that the scene in front of the imager remain static during the time required to capture the entire hyperspectral cube. The resulting images also involve huge amounts of data that must be subsequently transferred off the imager, often to another location for processing and interpretation. For these reasons, hyperspectral imaging is usually a good fit in agriculture only in crisis situations and some postharvest inspection scenarios. During fruit inspection, HSI has been used to successfully identify bruising; quantify firmness; and detect moisture, soluble solid content (SSC), acidity, starch, phenols, anthocyanins, lycopene, total chlorophyll, ascorbic acid, and other biochemical constituents. During safety inspection, HSI has also been used to detect insect damage, feces contamination, and fruit deterioration from a variety of causes (Pu et al. 2015). In the fields, hyperspectral imaging is an emerging technique in precision agriculture that is especially well suited for finding visual abnormalities that are detectable by the human eye. HSI is nondestructive, provides consistent results, and is highly accurate for monitoring nutrients in crops, detecting water stress, identifying disease and insect damage, and assessing overall plant health. However, since tomato color can be generally assessed using the human eye (or imagers that are designed to mimic the human eye, as is the case with conventional digital cameras), HSI for this purpose alone is unnecessary and cost prohibitive.

A less cumbersome but often equally effective alternative to hyperspectral imaging is multispectral imaging. Multispectral imaging evaluates light scattered from objects, including plants and crops at specific wavelengths rather than a whole spectrum of wavelengths, as is the case with hyperspectral imaging. Ideally, these limited and specific wavelengths of incident light (and resulting scattered light) are chosen to provide the greatest resolution and information for monitoring what is of interest. For tomatoes, specific wavelengths within the red and green bands of the visible spectrum (Figure 5.2a) can be chosen to optimize resolution and accuracy in monitoring ripeness and optimal harvesting times. Incident light wavelength can be controlled in two different ways, either by choosing a light source to cover the desired wavelength or by filtering a broad spectrum light source to select only the wavelengths of interest. The former method requires narrow-bandwidth light sources such as lasers, which can be expensive and limited in availability. The latter method uses filters, which, while less expensive than lasers, attenuate the intensity of the incident light source on the fruit and, as a result, can limit the accuracy of resulting scattered light measurements. Light-emitting diodes (LEDs) are another option for multispectral imaging, but LEDs have a broader color (wavelength) bandwidth than lasers and can limit color resolution for the purposes of tracking ripening on a day-to-day basis. LEDs also do not deliver the power and intensity that lasers do, which can be advantageous for decreasing power consumption on mobile devices but may not provide sufficient illumination for accurate multispectral images.

Multispectral imaging has been used to monitor firmness, SSC, and other properties in a wide range of fruits (Li et al. 2018, Lorente et al. 2012). For tomatoes in particular, multispectral imaging has been used to discriminate tomato varieties with over 85% accuracy, determine lycopene and phenolic content (Liu et al. 2015), identify insect frass (feces) on mature tomatoes to address food safety concerns (Yang et al. 2014), and distinguish green tomatoes destined to ripen postharvest from those that will never ripen (Hahn 2002).

For in-field mobile platform measurements, multispectral imaging is a promising technique because, unlike conventional two-dimensional imagers, it can be customized for critical wavelengths that are matched to the task at hand. In contrast to hyperspectral imaging, multispectral approaches are also lower cost, produce less data, and require less power and bandwidth to gain valuable insight about populations of tomatoes, whether on or off the field.

FLUORESCENCE

A fluorophore absorbs light to varying degrees, but only in certain wavelength bands, and emits light of a specific color in response. Fluorescence sensors, called fluorometers, exploit this property of certain natural substances such as chlorophyll by exciting the fluorophore with light it can absorb and measuring the fluoresced light in return. Chlorophyll content in a tomato declines during the ripening process. Chlorophyll is also a naturally occurring fluorophore, making chlorophyll concentration inside ripening fruit a possible candidate for sensing the perfect tomato. The fluorescent qualities of chlorophyll have been exploited in fruit-monitoring applications using actinic light sources to stimulate fluorescence in the fruit (Royer 1995). However, this method requires the fruit to be dark adapted prior to measurement and is therefore ill suited to measurements outside of the laboratory. Portable fluorescence monitors such as the Multiplex by Force-A have used multiple LED-based light sources to stimulate fluorescence associated with chlorophyll as well as anthocyanins and flavenols (Cerovic et al. 2009). These measurements do not require dark adaptation and are suited to portable field measurements. Even when not requiring dark adaptation, however, measuring fluorescence requires minimizing or eliminating ambient lighting to reduce interference and for this reason is inherently unsuitable for high-throughput monitoring in outdoor environments.

SEEKING THE COLOR OF THE PERFECT TOMATO IN THE INTERNET OF THINGS

Multiple color-sensing technologies are compatible with IOT-scale monitoring of tomato color (Table 5.1). Whether a particular technology is a good choice for individual growers depends on a variety of factors including the size of the farm and the end use of the tomatoes grown there. For experimental farms whose end goal is to provide modeling, research results, and other data individual farmers can leverage for greater yield, more precise benchtop equipment such as spectrometers can be appropriate. Benchtop spectrometer software is certainly powerful enough to process color spectra and broadcast that data in a usable form to farmers within a region. On the other end of the spectrum, small farms with fewer resources can benefit from several well-placed cameras in the fields or even from color sensors integrated into a wearable package for use during picking. Postharvest, cameras can support monitoring color throughout sorting, handling, transport, and distribution.

Accessibility (Table 5.2)

With the exception of custom fluorometers (such as those requiring dark adaptation and actinic light sources), color-sensing technologies suitable for monitoring tomatoes in the field are readily available commercially. Of these technologies, conventional imagers or cameras are most accessible to the individual farmer.

TABLE 5.1 Suitability of Color Monitoring Sensors to IOT Scale Sensing

Technology	Overall	Accessibility	Compatibility	Density	Performance	Versatility
				IOT Characteristic		
Spectrophotometers	Good	Fair	Fair	Poor	Very Good	Very Good
Colorimeters	Good	Good	Good	Poor	Good	Good
Color Sensors	Good	Good	Very Good	Good	Good	Good
Conventional Cameras	Very Good	Very Good	Very Good	Very Good	Good	Very Good
Hyperspectral Imagers	Fair	Poor	Poor	Good	Very Good	Good
Multispectral Imagers	Good	Fair	Good	Good	Very Good	Fair
Fluorometers (Actinic Light)	Poor	Poor	Poor	Poor	Good	Poor
Fluorometers (LED)	Fair	Fair	Fair	Fair	Good	Fair

TABLE 5.2   Accessibility of Color Sensing Technology

	Technology Characteristic				
	Affordability	Availability	Battery Life	Compactness	Ease of Use
Spectrophotometers	Low	High	Low	Low	Low
Colorimeters	Average	High	Average	Average	Average
Color Sensors	High	High	High	High	High
Conventional Cameras	High	High	High	Average	High
Hyperspectral Imagers	Low	Average	Low	Low	Low
Multispectral Imagers	Low	Average	Low	Low	Average
Fluorometers (Actinic Light)	Fair	Fair	Low	Low	Low
Fluorometers (LED)	Fair	Fair	Average	Average	Low

Imagers are made by a wide variety of manufacturers, enjoy the low cost benefit of high-volume global production, can operate on batteries, and have image compression algorithms (e.g., JPEG and others) readily available to reduce the bandwidth of data that needs to be transferred off the camera or sensor itself. Other technologies, such as the spectrophotometer, have been commercially available for several decades and have been used extensively in agriculture and plant research, but remain prohibitively expensive for many small- to medium-sized farms.

Compatibility (Table 5.3)

To be true to the vision and goals of the IOT, sensors should operate with as little user intervention as possible and be able to adjust automatically to changing conditions in the sensing environment while restricting data transfer to what is most relevant to a particular problem. Color sensors should be readily customized to particular problems and contain on-board intelligence to condition, preprocess, and compress data before transfer off the node to the cloud. Spectrophotometers, whether benchtop or handheld, have

TABLE 5.3 Compatibility of Color Sensing Technology with Sensor Platforms

| | Platform | | | | | | | |
| | Preharvest | | | | | Postharvest | | |
	Benchtop	Handheld	Stationary	Mobile	Wearable	Storage	Conveyer	Package
Spectrophotometers	✓						✓	
Colorimeters	✓	✓						
Color Sensors		✓	✓	✓	✓	✓	✓	✓
Conventional Imagers		✓	✓	✓	✓	✓	✓	✓
Hyperspectral Imagers				✓			✓	
Multispectral Imagers				✓				
Fluorometers (Actinic)	✓							
Fluorometers (LED)	✓	✓						

been customizable for many years. Wavelength range and resolution are adjusted to suit the customer and are typically an integral part of the purchasing and selection process. However, these instruments typically collect only raw data and rely on software to condition that data, often without significant compression or bandwidth reduction. Due to the fundamental premise on which spectrophotometers and fluorometers are designed, as instruments rather than sensors, these technologies are not particularly compatible with the IOT. Rather, other sensing technologies, such as cameras and color sensors, are much more autonomous and suitable for compressing data for transfer to the cloud in an application-specific and intelligent way. Combining camera capability with spectrophotometry, hyperspectral and multispectral imagers are more instrumental and less suited to a "smart" sensor approach, although scanning of a scene during the imaging process is highly automated to ensure that reliable data is collected.

To be useful for monitoring pre- and postharvest tomatoes, sensor approaches should also be compatible with typical farming operations, equipment, and infrastructure. Some approaches to color sensing are less suited to farms than others. For example, the benchtop spectrophotometer typically requires a laboratory-type space as well as consistent sampling protocols for best results. Farms, particularly individual ones in the small to medium range, rarely host such spaces. Other sensors (including portable spectrophotometers) are more portable and suitable to a broader range of platforms, many of which can provide the density and versatility necessary to support widespread collection of data for monitoring individual plants or tomatoes.

Conventional imagers or cameras are particularly compatible with multiple-location, on-field monitoring of tomato color throughout the crop maturation and ripening and harvesting process. Stationary, high-resolution monitoring reduces the complexity of tracking individual tomatoes by virtue of the fact that day to day, tomatoes on the vine remain in approximately the same location as they were the day before. Subsequent image processing is reduced because there is no need to search far and wide in a range of images to superpose images for the purpose of tracking ripening. And, conventional imagers are also useful for multiple tasks that extend beyond color monitoring. They can be used to detect weeds, track plant health including leaf chlorophyll content, and monitor ambient light exposure, thereby providing a comprehensive understanding of tomato crops at not only the individual tomato but the individual plant level. Similar benefits can be obtained from using cameras in conveyer belt operations used postharvest for either fresh or processing tomatoes. Machine vision and learning algorithms for image processing are highly advanced and can solve a myriad of problems associated with the health and well-being of tomatoes both pre- and postharvest.

Although further into the future, wearable sensors for monitoring color are also viable for monitoring tomato crops. Small individual or small arrays of color sensors integrated with electronics that deliver selective information regarding tomato color can vastly reduce the subjectivity and variability in the manual operations that continue to dominate fresh tomato growth and production. Such smart wearable sensors could be integrated into gloves or mounted on eyewear to support objective, reliable, and consistent evaluations of color for understanding crop health, harvest readiness, suitability for market, homogeneity, and other properties.

Density

IOT-scale sensors must be deployed at a large enough scale that the amount of data collected provides a high-resolution picture of the physical world. Collectively, the picture provided by IOT data should be much greater than what is currently possible with standalone sensor technologies. To achieve the density expected of IOT-scale sensing, color sensor technologies should be able to monitor the color of many tomatoes and do so both on a fruit-by-fruit basis and on a timescale approaching every 24 hours (daily). Several of the technologies discussed in this chapter have the ability to provide such high spatial and temporal resolution data. Unfortunately, traditional methods such as those involving sampling and processing a tomato in order to collect color data cannot be done at IOT-scale density. When these traditional techniques are miniaturized and power consumption reduced, they can be deployed in handheld spectrophotometers and colorimeters that can provide a much higher number of on-field measurements while also avoiding processing and destruction of the tomatoes tested. Still, handheld monitors require significant labor resources, which ultimately limits the number of color measurements and individual tomatoes that can be monitored in the field. The color-sensing technology used in handheld and benchtop colorimeters, however, is rapidly advancing, making more automated color detection in the field possible in the near future. Taking the human aspect out of color measurements decreases demand on labor resources and increases the total amount of data that can be collected. Even more promising, imagers used in commercial digital cameras can be mounted on stationary locations in a field (e.g., poles) or on manned (e.g., tractors) or autonomous (e.g., ground robots or drones) vehicles, providing almost unlimited color information on an RGB scale. While deployment of cameras on mobile platforms or vehicles greatly increases the signal processing and location accuracy needed to compare color changes for individual tomatoes during the ripening process, these mobile measurements are excellent for population assessments. The same is true for both hyperspectral and multispectral imaging, although the increased cost and power consumption of these technologies make installation on stationary platforms impractical.

Performance (Table 5.4)

The performance of color-sensing technologies depends largely on their design approach. Spectrophotometers are designed to measure color at very high resolution at a single point. Thus, their accuracy and color resolution are high while their spatial resolution remains low. Colorimeters and color sensors behave essentially as spectrophotometers, measuring color at a single point, albeit at much lower color (wavelength) resolution. Cameras, on the other hand, are designed to measure images, capturing color at high spatial

TABLE 5.4 Performance of Color Sensing Technology

				Performance Metric			
	Accuracy[a]	Durability	Dynamic Range	Color Resolution	Spatial Resolution	Stability[b]
Spectrophotometers	Average	Low	High	High	Low	High
Colorimeters	Low	Average	Average	Average	Low	High
Color Sensors	Low	High	Average	Low	Low	Average
Conventional Imagers	Low	Average	Average	Low	High	Average
Hyperspectral Imagers	High	Low	High	High	High	High
Multispectral Imagers	High	Low	High	Average	High	High
Fluorometers (Actinic)	Low	Low	High	Low	Low	Low
Fluorometers (LED)	Low	Low	High	Low	Low	Average

[a] Refers to a combination of how well the sensor can determine color in space (multiple dimensions) and across wavelength.
[b] Refers to vulnerability to photobleaching, ambient light interference, dark adapting, and other destabilizing factors.

resolutions but limited wavelength resolutions (red, green, blue). Hyperspectral and multispectral imagers combine qualities of both cameras and spectrophotometers, providing high color resolution at spatial resolutions that far exceed conventional spectrophotometers. And, finally, fluorometers are specialized instruments designed to measure light emitted from fluorophores like chlorophyll that are found naturally inside plants and fruits. For monitoring color in tomatoes, where a fluorometer would rely on natural fluorophores, accuracy and resolution can be quite low because the fluorescence signal involved is small and subject to interference. Outside of a highly controlled environment, fluorometers can have a wide range of performance issues. And, fluorometers as well as spectrophotometers will, in many cases, require glass optical fibers to attain maximum accuracy. This limits their durability, as glass cracks easily, particularly in field use situations.

Versatility

To be viable in IOT-scale agriculture, sensing technologies should also be versatile, amenable to reconfiguration that is consistent with redefining a sensing problem as it evolves in pre- and postharvest scenarios. Of the multiple sensing technologies available to farmers for monitoring color, colorimeters are the least versatile, as they provide only three axes of color related to overall luminance, blue/yellow, and red/green colors in an image. Spectrophotometers, while providing highly detailed color information, are typically shipped in a fixed configuration that is difficult if not impossible to change on-site. Affordable fluorometers are often made to suit a particular fluorophore (e.g., chlorophyll), compromising versatility for reduced cost. Hyperspectral and multispectral imagers are typically configured according to how they are deployed, thus limiting their ability to cross platforms from the ground to the air or vice versa. In contrast, conventional imagers can be adapted by on-board microcontrollers and signal processing to a wide variety of problems quite easily and are the most versatile of the color-sensing technologies considered herein.

CONCLUSIONS

Throughout the history of domesticated tomato farming and production, color has been important and often used to judge both the quality of individual tomatoes as well as their readiness for harvest. At a crop level, the distribution of color can be used to identify maturity, assess fruit and crop health, and estimate yields. Postharvest, color and changes in color over time are a primary indicator of ripening, over-ripening, and overall quality. Few would disagree that color provides valuable estimates of how good a tomato is today and how good it is likely to be tomorrow.

Fortunately, color sensors, ranging from single RGB sensors to multiple-wavelength, high-resolution hyperspectral imagers, are widely available commercially. Cost ranges from a few dollars to tens of thousands of dollars, and many color sensors, especially cameras, can be adapted to multiple problems in pre- and postharvest situations. Color sensing poses few barriers to entry for farms and processing operations, large and small. Through the interconnectedness and autonomy provided by the IOT, color sensors can provide a wealth of data and insight into optimizing crop yields that easily justifies their use, installation, and cost.

CHAPTER 6

Sensing for the Perfectly Firm Tomato

F IRMNESS IS A WIDELY accepted indicator of the ripeness of hard fruit like apples. It is also considered one of the best indicators of bruising and for this reason complements color measurements in the evaluation of fruit quality before, during, and after harvest (Valero et al. 2003). For tomatoes in particular, firmness is also a key predictor of culinary quality and shelf life (Wann 1996). Firmness progresses differently during ripening and maturation depending on tomato variety and is also influenced by chilling. Chilling occurs when ambient temperatures, while still above freezing, are low enough to compromise the health and quality of a maturing tomato (Jackman et al. 1990).

Fruit firmness is traditionally measured as the force required to induce tissue collapse. Tissue collapse is induced by and measured with an instrument called a penetrometer. Unfortunately, inducing tissue collapse requires puncturing the fruit and by its very nature often makes the tested fruit subsequently unsuitable for sale and consumption. Alternatives to traditional penetrometers are available that use this *force vs. deformation* approach to deduce firmness but create only microdeformations during the testing process and are minimally damaging. Other mechanical means for testing firmness are also available, as are nonmechanical approaches to sensing firmness, including acoustic, optical, and magnetic techniques. Still other approaches borrowed from tactile sensing in robotics can combine contact area and force sensing to deduce firmness and other mechanical properties. Each of these approaches relies on what are ultimately imperfect sensors, the trade-offs for which are discussed alongside fundamentals of sensor operation in this chapter.

TRADITIONAL APPROACHES TO MEASURING FIRMNESS

For many years, penetrometers (Figure 6.1a) have been the instrument of choice for measuring the firmness of fruits and vegetables. Penetrometers use the Magness-Taylor test, which requires manually peeling away part of the fruit, applying a cylindrical plunger to the peeled area, and measuring the

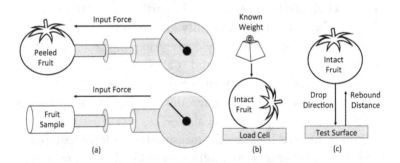

FIGURE 6.1 Approaches to detecting firmness using traditional force-based measurements. (a) The traditional penetrometer (Magness-Taylor) test measures the force required to puncture a peeled portion of fresh fruit or a cylindrical sample of the fruit; (b) an impact approach measures the amount of force and the duration (time) of impact between the fruit and another object (e.g., a load cell); and (c) the rebound technique measures the bounce-back (rebound) distance after the fruit is dropped from a known distance to a stationary surface.

maximum force required to penetrate the pulp. In a softer fruit like a tomato, puncture pressure is often measured as a function of applied force on an unpeeled, intact tomato. Measured in this way, tomato firmness can vary over threefold during the ripening process. Mature green tomatoes will deform about 0.5 mm at an applied force between 55 and 75 Newtons (N), while ripe tomatoes will require only 16–25 N to deform that same amount (Wann 1996). Firmness can also be measured by picking, sampling, and preparing a cylindrical sample of tomato with known geometry. The sample can then be punctured to the point of tissue collapse and the resulting force correlated to firmness. Using this approach, tissue collapse typically occurs at less than 10 N for ripe tomatoes and at around 50–60 N for mature green tomatoes (Jackman et al. 1990, Wann 1996). The penetrometer can be held by hand or mounted on a surface for greater accuracy. Different-sized tips are available for different types of fruit depending on their rupture pressure. The greater the rupture pressure, the larger the plunger tip. Commercial penetrometers are made by a variety of manufacturers, including AgTec, PCE Instruments, SI Instruments, TR Turoni, and Wagner Instruments, and range in price from $125 to $500.

Higher-resolution tests that provide multiple data points above and beyond puncture pressure can also be performed using the Magness-Taylor force vs. deformation test. This alternative test measures deformation as a function of applied pressure up to and including puncture. Regardless of whether the test measures only the rupture pressure or a pressure-deformation curve up to and including rupture, the Magness-Taylor test requires picking of the fruit, and the test itself often damages the fruit. Furthermore, measurements are often inconsistent and can vary up to 30% for fruit with identical properties (García-Ramos et al. 2005). Thus, these traditional techniques may not be well suited for collecting, comparing, and evaluating large quantities of firmness data collected at different times and in different places.

Alternatives to the Magness-Taylor test can be discussed from multiple perspectives. For the purposes of evaluating sensors required to support these alternative tests, this discussion looks at the four major types of signals that can be used to represent firmness: mechanical, optical, acoustic, and magnetic. Mechanical approaches apply a force, pressure, or other mechanical stimulus to the fruit and measure the deformation or change in another mechanical property of the fruit that results. Traditional methods use mechanical force vs. deformation, but *force vs. impact, force vs. rebound*, and *force vs. contact area* can also be used to evaluate firmness. Alternatively, optical, acoustic, and magnetic approaches send a signal into the fruit using relatively little contact and measure resulting changes in output light, vibration, and magnetic field respectively to deduce firmness. While mechanical approaches by their very nature are more closely connected to what is being measured (firmness) and tend to require less expensive sensors, most are also inherently damaging to the fruit. On the other hand, optical, acoustic, and magnetic sensors can operate without making damaging contact with the fruit but tend to require more expensive sensors or produce signals that are more vulnerable to interference caused by both ambient conditions and variations in other fruit characteristics.

MECHANICAL APPROACHES TO MECHANICAL APPROACHES TO MEASURING FIRMNESS

A number of different approaches have been used to measure the firmness of fruit through the application of a mechanical stimulus: force vs. deformation, force vs. impact, and force vs. rebound. Other approaches based on the human sense of touch that rely on force vs. contact area may also be used to measuring the firmness of a tomato.

Force vs. Deformation: Consistent with the Magness-Taylor test, the most common approach for mechanically evaluating firmness is to apply a force or pressure and measure the resulting deformation or rupture pressure of the fruit. In a tomato, the deformation that occurs in response to an applied force or pressure provides an estimate of firmness whether or not the tomato is peeled. This approach is also the most common manual technique consumers use to evaluate firmness. By squeezing the fruit without destroying or puncturing the skin, the typical consumer gains a reasonable idea of how firm a tomato is and, by extension, how ripe or how damaged it is.

A probe in the shape of a sphere that is pressed against the surface of a tomato with a force F produces a deformation d according to Hertz's contact stress theory:

$$F = \frac{4}{3} E^* R^{0.5} d^{1.5} \tag{6.1}$$

where R is the radius of the sphere and E^* is related to the elastic properties and can be directly correlated to the firmness of both the tomato and the spherical probe through:

$$\frac{1}{E^*} = \frac{1-\nu_1}{E_1} + \frac{1-\nu_2}{E_2} \tag{6.2}$$

where E_1 and E_2 are the elastic moduli of the tomato and the probe, respectively, and ν_1 and ν_2 are the Poisson's ratios of the tomato and probe, respectively. If the probe is made of a hard material that deforms little relative to the tomato, $E_2 = \infty$, and this equation simplifies to:

$$\frac{1}{E^*} = \frac{1-\nu_1}{E_1} \tag{6.3}$$

In theory, a nondestructive force or pressure large enough to measure E^* but not so large as to cause damage can be applied to the tomato for a fixed period of time and the resulting deformation measured by an appropriate sensor. In practice, however, E^* changes with applied pressure, and measurements of E^* at low applied forces do not correlate well with human perceptions of fruit firmness. To achieve firmness measurements that coincide with these perceptions, applied forces must often approach a level that causes bruising to the fruit (Pitts et al. 1997). In order to avoid irreparable damage, the applied force must be carefully controlled to be both relevant to firmness and to stop short of damaging the tomato under test. This can be done either by applying a force that remains beneath the damage threshold and measuring the resulting deformation, or by creating a controlled but not permanently damaging deformation or dent in the fruit surface and measuring the resulting force. In either case, the input stimulus must be carefully controlled to avoid permanent damage to the fruit.

A number of devices and instruments have been developed to exploit this nondestructive microdeformation approach to determining firmness (García-Ramos et al. 2005). For example, one instrument uses a spherical probe attached to a spring to push against the side of the fruit through a soft spherical cup until a maximum deformation of 2 mm is achieved. The compression of the spring indicates firmness of the fruit without damaging it. Alternatively, the Durofel instrument forces a flat-ended rather than spherical probe into the fruit and uses either an analog spring and gauge or digital sensor readout to indicate resulting hardness or firmness. While this device has been extensively used to gauge firmness in firm fruits such as apples, the probes are still quite large, with contact areas ranging from 10 to 50 cm^2, and are still likely to bruise soft fruit. The Commonwealth Scientific and Industrial Research Organisation (CSIRO) uses a different approach that requires placing a fruit into a v-shaped clamp, putting a weight on it, and measuring the resulting displacement after a certain period of time. This approach, while less likely to bruise soft fruit than the Durofel device, requires both picking and careful placement of the fruit, which may limit its application to measurements where the value of quantifying firmness clearly justifies the time taken to measure it.

Force vs. Impact (Figure 6.1b): When a fruit, such as a tomato, collides with another elastic object, the force and duration related to the impact are directly related to the firmness of the fruit (Chen et al. 1996, García-Ramos et al. 2005). There are two basic ways to exploit this behavior for sensing fruit firmness: dropping the fruit onto a sensor or dropping a sensor onto the fruit. In the former scenario, a known weight is typically dropped onto a fruit as it sits on or is otherwise attached to a load cell. The load cell contains a sensor or sensors that monitor the force during impact, and variations in force over the duration of impact are used to determine fruit firmness. This method, while nondestructive and minimally bruising, typically requires the fruit to be picked prior to testing, which can limit its use for understanding the quality of individual tomatoes. This method also requires precise placement onto the load cell, which may incur additional overhead and cost in postharvest applications.

Rather than dropping a known weight onto the fruit and relying on gravity to generate force vs. impact information, another implementation of this approach uses a mechanical probe to tap the fruit in a controlled manner. During the controlled (input) stimulus, force is monitored and used to determine firmness. Some instruments using this force vs. impact technique have been commercialized for use in production lines. For example, the Sinclair iQ tester has a pneumatically operated sensor head that is used to tap a fruit using a constant initial force. The force experienced when the sensor head collides with the fruit during this nondestructive tap is used to compute firmness using the Sinclair firmness index, which consists of a scale of 0–100 arbitrary units (Valero et al. 2003). The Sinclair iQ approach does not necessarily require a fruit to be harvested prior to testing and has been successfully demonstrated for in-situ testing in apple and pear orchards (Valero et al. 2003). This method also produces acceptable repeatability for tomato firmness testing (De Ketelaere et al. 2006). Other instruments also implement force vs. impact, but most are still used only for research purposes, and those that are commercialized use proprietary technology where underlying sensing mechanisms are not readily available (García-Ramos et al.

2005). Furthermore, unlike the force vs. deformation method, using the force vs. impact method is inherently a two-step process requiring the application of a force in one step and the response of the sensor when it makes contact with the probe in the next. Two-step measurements increase the measurement time, complexity, power consumption, and cost over one-step measurements.

Force vs. Rebound (Figure 6.1c): This category of mechanical approaches to measuring firmness also involves exploiting the elasticity of fruit. Like force vs. impact, this approach can be implemented using one of two basic strategies, both of which are designed to create a collision between the fruit and another surface or object. In the first, an elastic object collides with the fruit and the object bounces off the fruit, and in the second, the fruit collides with an elastic surface and rebounds or bounces off that surface. In both cases, the force of the collision is known and the rebound distance is used as a measurement of firmness. Olives have been dropped onto an elastic surface on production lines in order to successfully determine their firmness and quality (Barreiro et al. 2002). Unfortunately, tomatoes are too fragile to make this technique a viable one for identifying their firmness. And, while it is possible in principle to repel objects with a known force or weight off the side of a tomato and measure the resulting rebound behavior, doing so incurs a level of complexity and control that is not competitive with force vs. deformation or force vs. impact approaches to monitoring firmness.

Force vs. contact area (Figure 6.2): This approach to evaluating firmness is based on the human sense of touch and has been of great interest to the development and advancement of robotic hands and other robotic devices. In mimicking the sense of touch, force is monitored simultaneously with contact area to deduce firmness. The more contact sensors that are activated for a given force, the less firm the object. The tactile sensing approach is enormously useful for enabling robotic devices to grasp an unfamiliar object well enough to avoid dropping the object but not so firmly as to break it. This approach is also consistent with

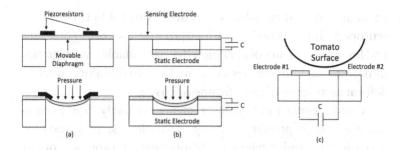

FIGURE 6.2 Alternative approaches to detecting firmness. A promising alternative to traditional mechanical means to measure tomato firmness is the tactile sensor, which, rather than relying on the relationship between two parameters at a single point in the fruit (e.g., force vs. deformation, force vs. time, force vs. rebound distance), uses the relative distribution of force over a multipoint contact area to deduce firmness (similar to how the human hand senses firmness). Particularly low-cost tactile sensors are those based on (a) piezoresistive sensors; (b) capacitive sensing technology to measure pressure; and (c) capacitive touch sensors to measure contact area.

nondestructive testing where the goal is to apply enough force to estimate the firmness of a fruit but not too much pressure that the fruit is irreparably damaged.

Unlike most other mechanical approaches to determining fruit firmness, tactile sensing is well suited to seamless, opportunistic sensing of tomatoes throughout maturation, harvesting, and postharvest handling and transport. Whether installed in mechanical harvesters that dominate the tomato-processing scene or integrated into wearable platforms for the hand harvesting of fresh tomatoes, tactile sensors can provide fast, unbiased evaluations of firmness prior to and during harvest. Furthermore, advances in tactile sensing technology will continue to be driven quickly by demand for such sensors in the various industries served by robotics: health care, automotive, manufacturing, mining, military, public safety, and others. Zou et al. (2017) provide an overview of both tactile sensors suitable to robotic and wearable platforms and smart sensor approaches to tactile sensing. Of particular interest are advances in inkjet-printed touch or tactile sensors (Salim and Lim 2017), which promise low-cost, lightweight, and low-power modes of operation that are suitable for integration into wearable or robotic devices as well as into traditional force vs. deformation approaches to firmness sensing.

Force vs. contact area measurements are likely to use one or more of three primary sensing technologies: piezoresistive, piezoelectric, and capacitive. Piezoresistive sensors are one of the simplest means to transform mechanical forces into signals useful for estimating firmness (Doll and Pruitt 2013). These sensors respond to applied stress, pressure, force, or touch with a change in electrical resistance. When a material such as silicon, germanium, or diamond (or a structure such as a carbon nanotube or carbon nanowire) mechanically bends or deforms in response to an applied force (Barlian et al. 2009), resistance changes either isotropically (i.e., the same in all directions) or anisotropically. Wheatstone bridge measurement circuits can be used to eliminate the large baselines in these piezoresistors and provide an output that is nonzero only when the sensor is perturbed (i.e., measuring an input of interest). Piezoresistors are small and well suited to being implemented in high-resolution arrays for accurate contact area information in small spaces. Furthermore, because many piezoresitive materials can be used in standard integrated circuit fabrication processes, piezoresistive sensors can also be easily integrated with readout and processing circuitry, thus reducing overall cost and size. For their size and ease of fabrication, these sensors are also quite robust, except for inherent hysteresis, which may limit their use as force sensors. Hysteresis will produce inconsistency in measurements between squeezing the target fruit compared to releasing pressure or force on that same fruit.

Piezoelectric sensors are similar to piezoresistive sensors, but they convert an input force or pressure to an electrical voltage rather than a resistance. The application of a force to a piezoelectric sensor polarizes the piezoelectric material, creating a separation of charges and an electric potential. This separation of charges is typically measured by placing the piezoelectric material in a circuit, which creates an electrical oscillation whose frequency varies with the applied force. Like piezoresistive sensors, piezoelectric sensors can be readily integrated with interface circuits to make small compact systems. However, the oscillator circuits required to actively control sensor behavior increase the size and power demand over piezoresistive sensing systems. Piezoelectric sensors are also not well suited to measuring static forces or pressures. Only when firmness changes can they provide an accurate reading. In combination, these drawbacks often limit the application and competitiveness of piezoelectric sensors for measuring force, pressure, and contact (touch).

Finally, the last primary option for measuring both force and contact (touch) is the capacitive sensor. These sensors are widely used for touch screens in smartphones, tablets, and other consumer electronic devices. They operate on the principle that the distance between two conductive materials separated by an insulator or the overlapping area of the conductive materials will change, thus changing the capacitance as a function of incoming force, pressure, contact/touch, or acceleration. Because capacitive techniques require more sophisticated readout circuitry than piezoresistive sensors, they are often more complex, larger, and more power hungry. However, they do not exhibit the increased electronic noise that is inherent to resistors, nor do they demonstrate hysteresis, thus allowing greater repeatability and reliability in firmness detection applications.

For sensing the perfect tomato, whether by a human being who is manually grasping the fruit to assess firmness or by a futuristic robot performing more automated assessment, these three technologies are promising. Although each technique has performance advantages as well as drawbacks (Table 6.1), piezoresistive sensors are likely the best choice for measuring force (Almassri et al. 2013), while capacitive touch sensors may be the best choice for detecting contact area. Together, these two low-cost sensing technologies can simultaneously measure force and

TABLE 6.1 Performance Limitations of Firmness Microsensor Technologies

	Sensor Operation		Performance		Drawbacks			
	Input	Output	Sensitivity	Resolution	Hysteresis	Crosstalk	Drift	Temp Sensitive
Piezoresistive	Pressure	Resistance	Moderate	Moderate	Yes	No	Yes	No
Piezoelectric	Pressure	Voltage	High	High	Yes	No	No	Yes
Capacitive	Pressure	Capacitance	Moderate	High	No	Yes	No	No

contact area to deduce firmness similar to the way the consumer does it—by touch.

OPTICAL SYSTEMS FOR FIRMNESS MEASUREMENT

As an alternative to force, an optical signal (light) can also be used to measure the firmness of tomatoes. Two primary optical approaches are available. The first is based on the force vs. deformation method and uses optical rather than mechanical sensors to measure the deformation that results from an applied force. The second method is both less direct and less invasive and uses both an optical input and output signal to capture certain properties of fruit that have a known relationship to firmness.

In the first method, an optical waveguide in the shape of a fiber, rectangle, or other structure is applied to the tomato with some pressure. An optical waveguide is a structure that restricts where and how light can travel. Changes in the shape of the waveguide, through contact with the fruit surface, cause a change in how light propagates or travels through the waveguide. Changes in light transmission characteristics can mean a change in phase, wavelength, polarization, or intensity of the light traveling through the waveguide. In turn, these changes in light properties are proportional to sheer and vertical stresses on the waveguide as well as other mechanical properties that are a direct indication of firmness. Using an optical waveguide and sensor to determine firmness through the force vs. deformation approach provides some advantages over mechanical sensors. An optical output is inherently immune to electronic noise and can provide more consistent and reliable readings. However, an optical output also requires that the fruit under evaluation be antireflective and that interference from ambient light sources be kept to a minimum. While these two issues are of concern for any optical sensor or sensing system, they are less of a concern with the optical force vs. deformation approach because the waveguide does not depend on propagation of light through the ambient environment to provide an accurate indication of firmness.

Other optical methods use light both as an input stimulus and as an output signal. For example, near-infrared light (with wavelengths starting at 750 nm) can be used as an input stimulus and directed toward the surface of the fruit in a controlled manner and at a known angle. Some light is reflected off the surface of the fruit, while other light is scattered inside the fruit. Either scattered or reflected light can then be measured at known angles to the incident light, providing useful information regarding internal defects and fruit characteristics, including firmness. This optical approach is entirely noninvasive and nondestructive and has been shown to have reasonable correlation with firmness measurements collected using more traditional penetrometer-based techniques (Butz et al. 2005). Near-infrared (NIR) techniques (i.e., optical spectroscopy) have been used to estimate the internal sugar content and acidity of peaches, pears, kiwi fruits, and apples with reasonable accuracy. Using these NIR methods to measure firmness, however, has produced mixed results when compared to the Magness-Taylor test. Other optical methods using this spectroscopic approach to capture scattering of light from fruit have used lasers at certain visible frequencies (between 600 and 700 nm). Laser-based approaches can roughly categorize fruit by firmness, but results are still less accurate than those obtained using the standard Magness-Taylor pressure test (García-Ramos et al. 2005).

ACOUSTIC SENSORS AND SYSTEMS FOR MEASURING FIRMNESS

In the previous section, an optical method of detecting firmness using a mechanical input stimulus (i.e., a deformation) and an optical output was discussed. Acoustic sensing strategies are similar to this optical approach in that they use a mechanical stimulus (e.g., tapping the fruit) to generate the input stimulus. However, rather than reading the output as light, acoustic approaches read the fruit's response to this tapping by monitoring the propagation of the resulting vibration through the fruit in the audible range (0–20,000 Hz). Acoustic techniques fundamentally measure the modulus of elasticity (E) of the fruit by monitoring the resonant frequency of the vibration that moves through the fruit as a result of the input tap (García-Ramos et al. 2005):

$$E = Cf^2 m^{2/3} \rho^{1/3}$$

where the firmness index is approximately equal to $f^2 m^{2/3} \rho^{1/3}$ or variations thereof, as described in Cherng (2008), ρ is the density of the fruit; m is the mass of the fruit; f is the frequency of the maximum amplitude at the output signal (i.e., the resonant frequency) and C is a constant. This relationship allows the firmness of the fruit to be directly computed from the resonant frequency of the output signal if the mass and density of the fruit are known. The most common method of applying an input stimulus to the fruit in order to measure its vibration characteristics is to stimulate it in such a way that vibrations of multiple frequencies travel from one end of the fruit to the other. Sensors in contact with the fruit at the opposite end capture the vibration, and a Fourier transform (or similar signal processing technique) is used to divide the signal into its component frequencies. These acoustic signals can be captured using piezoelectric sensors that transduce the mechanical vibration in the fruit into an electrical signal in a manner similar to how microphones operate. The frequency with the largest amplitude in the resulting electrical signal is the resonant frequency and can be used to compute the firmness index as well as the elastic modulus of the fruit. This approach to evaluating fruit has been demonstrated in multiple instruments. AWETA makes an acoustical firmness sensor for laboratory use that mechanically generates a vibration at the top of the fruit and measures the transmitted vibration at the bottom of the fruit using an accelerometer. Similarly, the Firmalon (Shmulevich et al. 1996) uses three small hammers to generate an electromechanical impulse and three piezoelectric sensors to measure the resulting vibration after it has traveled through the fruit. Firmness is estimated using a selective average of the three sensor responses.

An alternative to using a mechanical input stimulus in acoustic approaches is to generate a controlled vibration using a piezoelectric (electrical to mechanical) vibration generator rather than using a purely mechanical technique—tapping the fruit. The piezoelectric material acts opposite to the piezoelectric sensor used to measure the output signal and generates an oscillating mechanical vibration in response to an oscillating electrical input. This signal is generated at a known frequency close to or at the resonant frequency, and the attenuation of the signal as it travels through the fruit is used to determine the modulus of elasticity and then the firmness index of the fruit. The advantage of using a piezoelectric generator is that the input signal can be better controlled than with a mechanical input stimulus, but a major drawback is that the piezoelectric signal is not very large and can attenuate or diminish below the noise floor of the corresponding sensor as it travels through the fruit under test.

Ultrasounds (frequencies greater than 20,000 Hz) are popular and highly effective for acoustic measurement systems and have a number of advantages over using signals in the audible range. Unfortunately, ultrasonic vibrations attenuate (diminish) quickly as they travel through plant tissue, and for this reason do not make good candidates for measuring the mechanical properties of tomatoes, including but not limited to firmness. Thus, acoustic stimulus and sensing techniques must rely on signals in the audible range when applied to evaluating fruit.

MAGNETIC SENSORS FOR MEASURING FIRMNESS

Magnetic fields can be manipulated to examine the detailed internal characteristics of everything ranging from the human body to fruit to unknown compounds. The most popular means of using magnetic fields to understand what cannot be seen or observed is nuclear magnetic resonance (NMR), which is often called magnetic resonance imaging (MRI) when applied to humans or animals. This technique capitalizes on the fact that hydrogen has a nucleus with a magnetic moment that can be examined in the presence of a strong magnetic field and certain radio frequencies. NMR can detect the presence of hydrogen under these conditions. Of most interest to analyzing food quality is 1H MRI, where a certain radio frequency is rotated in a magnetic moment and then removed. The resulting energy relaxation and energy loss are then measured and analyzed to identify water content and distribution in food, including fruits. Water content is directly related to fruit firmness.

NMR has the unique advantage of being able to identify the distribution of water inside a fruit without damaging it. Thus, it provides rich, high-resolution information about internal defects, water transport, and abnormalities as well as more general information that can be used to understand ripeness, decay, firmness, SSC (soluble solids content), and other parameters of interest both before and after harvest (Butz et al. 2005). NMR has been successfully used to identify defects and analyze the internal contents of a wide variety of fruit including avocados, peaches, nectarines, apples, and melons (García-Ramos et al. 2005), as well as tomatoes (Ciampa et al. 2010).

Until recently, however, NMR/MRI was out of reach to most farmers, requiring high magnetic fields and complex electronics that make these instruments prohibitively expensive for agriculture. However, lower-cost, lower-field MRI instruments have started to emerge in commercial markets. These instruments provide handheld, portable capability for monitoring food quality postharvest and possibly crop health in fields before harvest, although response times and processing speeds are still low compared to other techniques (Butz et al. 2005).

SEEKING THE FIRMNESS OF THE PERFECT TOMATO IN THE INTERNET OF THINGS

Firmness and fruit texture are critically important to the economic viability of tomato crops. Firmness influences consumer perception of tomato quality, indicates chilling damage incurred by exposure to low but not freezing temperatures, interferes with perceived aroma and taste, and affects shelf life and transportability (Bertin and Génard

2018, Jackman et al. 1990). To date, the preferred ways to measure firmness have been compression and puncture tests. These methods, while they agree well with human sensory evaluation (Causse et al. 2002), are often destructive or damaging to the tomato under test.

A variety of nondestructive methods for evaluating tomato firmness have emerged in the quest for alternatives to traditional techniques. The technologies best suited to evaluating firmness as discussed in this chapter are summarized in terms of their suitability to IOT-scale sensing in Table 6.2. All of these technologies can support farms that grow tomatoes for fresh consumption or those that grow the fruit for processed products, in open fields or in greenhouses. However, when considering the highly limited resources that small- and medium-sized farms often have to invest in technology as well as the large amount of data that is expected of IOT solutions, certain sensor technologies rise to the top in terms of combined accessibility, autonomy, compatibility, density, and versatility. Although not yet proven in agricultural settings, piezoresistive and capacitive sensors for biologically inspired (human touch) sensing may be a promising option for measuring the firmness of large populations of tomatoes, both before and after harvest.

Accessibility (Table 6.3)

Technologies designed to measure tomato fruit firmness vary widely in their accessibility to farmers and others involved in tomato production. Accessibility includes considerations of how readily available sensing technologies or instruments are on commercial markets and how much they cost (affordability) as well as their complexity (including maintenance), ease of use, power consumption and battery lifetime, and compactness (size and weight).

Traditional penetrometers that use force vs. deformation approaches to measuring firmness are readily available commercially and are fast and easy to use, making them accessible to a wide range of growers and farming operations. They are of moderate weight and size and can be used as handheld or benchtop instruments at moderate cost and low power consumption.

TABLE 6.2 Suitability of Firmness Sensors to IOT Scale Sensing

			IOT Characteristic			
	Overall	Accessibility	Compatibility	Density	Performance	Versatility
Macroforce Penetrometers	Low	Very Good	Low	Low	Fair	Low
Microforce Penetrometers	Moderate	Very Good	Low	Low	Good	Moderate
Force vs. Impact Instruments	Low	Fair	Low	Low	Fair	Low
Force vs. Rebound Instruments	Low	Fair	Low	Low	Fair	Low
Piezoresistive Sensors	High	Moderate	High	High	Good	High
Piezoelectric Sensors	Moderate	Moderate	Moderate	High	Good	Moderate
Capacitive Sensors	High	Moderate	High	High	Good	High
Acoustic Instruments	Low	Fair	Low	Low	Fair	Low
Magnetic Instruments (NMR/MRI)	Low	Fair	Low	Low	Very Good	High
Optical (Light Scattering)	Low	Low	Low	Low	Very Good	Moderate
Optical (Waveguide Deformation)	Moderate	Moderate	Moderate	Low	Good	High

TABLE 6.3 Accessibility of Firmness Sensing Technology

	Technology Characteristic					
	Affordability	Availability	Battery Life	Compactness	Simplicity	Ease of Use
Macroforce Penetrometers	Average	High	Average	Average	Average	Average
Microforce Penetrometers	Average	Average	Average	Average	Average	Average
Force vs. Impact Instruments	Average	Low	Average	Average	Average	Average
Force vs. Rebound Instruments	Average	Low	Average	Average	Average	Average
Piezoresistive Sensors	High	High	High	High	High	High
Piezoelectric Sensors	Average	Average	Moderate	High	Average	Average
Capacitive Sensors	Low	High	High	High	High	High
Acoustic Instruments	Average	Average	Moderate	High	Average	Average
Magnetic Instruments (NMR)	Low	High	Low	Low	Low	Low
Optical (Light Scattering)	Low	High	Low	Low	Low	Low
Optical (Deformation)	Low	High	Low	Low	Average	Low

However, the primary downfall of traditional penetrometers in terms of accessibility for IOT solutions is that they cannot be readily integrated with other sensors into arrays. Penetrometers tend to stand alone as instruments rather than sensors and thus have limited opportunities to become a part of a larger system. Other mechanical approaches to measuring firmness, including force vs. microdeformation, force vs. impact and force vs. rebound, tend to suffer from similar limitations.

In contrast to these traditional mechanical approaches, piezoresistive and capacitive sensors are easily integrated into arrays of like (homogenous) and unlike (heterogeneous) sensors. These sensors are also compatible with integrated circuit fabrication processes, which opens up the prospects for low-cost, and highly integrated, smart sensor arrays and systems. These sensors can be applied to force vs. deformation, force vs. impact, or force vs. rebound systems, or they can be designed into more biologically inspired systems that rely on detecting both applied force and contact area to mimic the human sense of touch. Because they offer similar functionality but have some additional drawbacks, piezoelectric sensors are not as attractive as piezoresistive and capacitive microsensors for capturing fruit firmness.

In addition to being able to measure force and pressure as standalone sensors, piezoelectric sensors are also a key part of acoustic sensing systems for measuring fruit firmness. Like piezoresistive and capacitive technologies, these sensors are also readily available commercially, although they have higher overhead in cost, power, and size because they require oscillator circuits to operate properly and piezoelectric crystals are often not compatible with integrated circuit fabrication. Unfortunately, many novel mechanical and acoustic sensing designs remain available only as research prototypes, which limits their availability to many potential users, including farmers.

Optical techniques, while available commercially as general-purpose instruments, are often larger than penetrometers and are quite expensive (ranging into the thousands of dollars). These two limitations alone limit their accessibility. Similarly, instruments based on NMR, while delivering unparalleled accuracy and detail regarding the internal defects and characteristics of fruit, remain far too expensive and complex to be accessible to most tomato farmers.

Compatibility (Table 6.4)

In addition to being accessible to tomato farmers, a sensor or sensor system must be largely or fully autonomous to be compatible with an IOT approach to sensing. Many traditional approaches to sensing firmness among tomatoes and related fruit do not fit this bill because they are often operated as handheld instruments rather than integrated sensors. This lack of autonomy applies to most instruments that require a mechanical input stimulus, including force vs. deformation, force vs. impact, and force vs. rebound, as well as acoustic sensing systems. It is worth noting that some of the commercially available instruments that use these techniques have digital displays and microcontroller-based interfaces, which make seamless, wireless transfer of instrument data to other devices or to more central locations in the cloud much more feasible. But, whether or not they create noticeable deformations during testing, most penetrometers and related instruments still require major user involvement in processing or placing the fruit and introducing a puncturing or penetrating force to the sample. Automating these processes is likely to require significant investment in robotics to mimic the capability of a human instrument user. Optical methods that rely on physical deformation of a waveguide suffer from similar limitations in autonomy and therefore are not particularly compatible with the IOT.

Other optical methods that rely on light scattering have more potential to be autonomous both at the data collection and data transfer levels. The proper projection of a light beam onto a tomato while it is still on the vine is fraught with complications associated with alignment and interference from ambient light. However, postharvest, when tomatoes are on a conveyer belt or similar

TABLE 6.4 Compatibility of Firmness Sensing Technology with Sensor Platforms

	Platform							
	Preharvest				Postharvest			
	Benchtop	Handheld	Stationary	Mobile	Wearable	Storage	Conveyer	Package
Macroforce Penetrometers	✓	✓						
Microforce Penetrometers	✓	✓						
Force vs. Impact Instruments	✓	✓						
Force vs. Rebound Instruments	✓	✓						
Piezoresistive Sensors					✓	✓	✓	✓
Piezoelectric Sensors					✓	✓	✓	✓
Capacitive Sensors					✓	✓	✓	✓
Acoustic Instruments	✓	✓					✓	
Magnetic Instruments (NMR, MRI)	✓					✓		
Optical (Light Scattering)	✓	✓					✓	
Optical (Waveguide Deformation)	✓	✓					✓	

production line, alignment, access, and interference become less serious issues, thus enabling scattering measurements to be collected, transferred, and managed without significant human oversight and intervention.

Magnetic methods, even when integrated into a portable MRI instrument, have similar user overhead to penetrometers and mechanical approaches to firmness sensing. While data can be collected, aggregated, and transferred autonomously from these sophisticated instruments, positioning the instrument to capture data is likely to involve user intervention into the foreseeable future, thus limiting their compatibility with the IOT.

Truly "smart" sensors consistent with the world of IOT are likely to be integrated packages of microsensor and electronics designed to produce firmness data with little to no intervention from the user before, during, or after measurement. Piezoresistive and capacitive sensors have already been integrated into many wearable device and robotic platforms to perform functions autonomously and seamlessly. Force vs. contact area measurements enabled by these two sensor technologies can also eliminate the need for a separate input stimulus, as they can be integrated into the normal handling of fruit before, during, or immediately after harvest. They can also be integrated into storage bins in order to monitor and limit the forces experienced by tomatoes in a way that reduces or eliminates postharvest bruising and other damage.

The compatibility of firmness sensors with existing tomato farming and production infrastructure must also be considered in judging overall suitability for the IOT. How well sensors and sensing technologies can be effectively integrated into existing operations platforms, including but not limited to automated vehicles (e.g., robots, drones), tractors, gloves, conveyor belts, and storage bins, is important. Unfortunately, most firmness sensor technologies are inherently incompatible with many of these platforms because they require contact with the fruit/tomato to make accurate measurements. All of the traditional mechanical approaches to firmness detection are subject to this limitation.

These technologies are entirely incompatible with aerial vehicles and most other mobile sensing platforms, with the exception of advanced ground robots, which might travel the fields grasping tomatoes, one by one, on farms in a more distant future. Traditional firmness measurement instruments are also largely incompatible with integration onto tractors, conveyer belts, and storage bins for the same reasons.

On the other end of the spectrum, noncontact technologies, such as those based on optical or magnetic (NMR or MRI) technologies, can potentially be put on any sensing platform that is large enough to carry the instrument. However, many of these complex systems remain compatible only with stationary or benchtop platforms because of their relative bulk and weight as well as the complexity and need for user setup and maintenance.

In sharp contrast to the limited sensing platforms that can host optical, magnetic, and penetrometer technologies, tactile (force vs. contact area) microsensor technologies are compatible with many farming platforms. As with most other firmness sensing technologies, the need for contact or near contact makes these sensors wholly incompatible with aerial vehicles and air-based sensing platforms as well as all but the most sophisticated ground robot mobile platforms. However, force vs. contact area sensors are compatible with many platforms that come into contact with tomatoes on a daily basis including storage bins, conveyer belts, and gloves. Sensor-equipped gloves can be useful when worn by harvesters who are picking tomatoes grown for fresh consumption by hand or for those on board mechanized harvesters who are assessing and sorting processing tomatoes.

Density

Since sensing firmness requires being physically close to or in contact with a tomato, technologies that can generate the dense, high-resolution data that is often expected of IOT solutions tend to also be those that are more autonomous and able to take data without user supervision. Preharvest, this means one of two things. First, wearable sensors (i.e., piezoresistive or capacitive) can collect data on large numbers of tomatoes while they are manually handled. An alternative to monitoring large numbers of tomatoes for their preharvest firmness is to use more complex instruments (i.e., optical and to a lesser degree magnetic) that collect large amounts of data on a much smaller number of tomatoes either using handheld instruments or off-field benchtop instruments. More dense data collected on a more limited number of samples is better suited to experimental farms or agricultural research to understand crisis scenarios or to optimize crop health on a much larger scale than individual farms. While both approaches are presently viable, the former uses large numbers of seamless sensors integrated onto wearable devices (e.g., gloves) and is more consistent with the spirit of IOT than the use of more complex instruments. While awaiting inexpensive futuristic robotic technologies that can handle and sense tomatoes automatically while they are still on the vine, the best choice for IOT-scale firmness sensing in terms of ensuring dense data may be those technologies that can piggyback on existing human activities and practices in the fields.

Postharvest, dense information can be collected using a larger range of options. For example, pressure, force, or contact area sensors can be integrated into storage bins to monitor or prevent batches of tomatoes from being subject to forces consistent with bruising or other damage. In a conveyer belt scenario where more precise positioning of the fruit is possible, sensors ranging from traditional penetrometer architectures to optical scanning can be used to capture firmness data from individual tomatoes, the percentage of which is dictated by throughput requirements. Technologies that require more precise placement, such as those that rely on providing a mechanical stimulus to the fruit (e.g., force vs. deformation, force vs. impact, force vs. rebound, and acoustic sensing technologies) and optical technologies that collect scattered light based on a controlled input light beam may be more expensive to implement than others. But, once implemented, these technologies can provide massive amounts of information regarding the firmness as well as other relevant mechanical properties of the fruit. In an IOT world, such dense data can be collected, aggregated, and processed fast enough to improve production handling of fresh tomatoes in real time and also aggregated with data collected from other operations to improve handling practices in the future.

Performance (Tables 6.5 through 6.7)

The performance of firmness sensor technologies varies widely. The greatest downfall of traditional technologies (penetrometers) is inconsistency (poor repeatability) caused by wide variations in user technique. Other mechanical approaches (force vs. impact and force vs. rebound), as well as acoustic instruments that rely on the manual application of an input force, are also vulnerable to human error and variation in technique, thus limiting both their accuracy and stability (including repeatability). Many penetrometers are also designed to detect force only at the point of tissue collapse, thereby limiting their dynamic range, although digitized versions and microforce penetrometers do offer broader force measurements. Thus, penetrometers and related firmness sensing techniques are best at classifying fruits into a limited number of categories of firmness but are often not well suited to continuous, high-resolution measurements of force.

TABLE 6.5 Performance of Traditional Firmness Sensing Technology

	Performance Metric			
	Classification Accuracy[a]	Durability	Dynamic Range	Stability
Macroforce Penetrometers	Low	High	High	Low
Microforce Penetrometers	Moderate	Moderate	High	Low
Force vs. Impact Instruments	Low	High	High	Low
Force vs. Rebound Instruments	N/A	High	Moderate	Low
Acoustic Instruments	Low	High	Low	Low

[a] Classification error reflect limits in both accuracy and resolution (Steinmetz et al. 1996).

TABLE 6.6   Performance of Force/Pressure Sensing Technology

	Performance Metric				
	Accuracy	Durability	Dynamic Range	Resolution	Stability
Piezoresistive Sensors	High	High	High	High	Moderate
Piezoelectric Sensors	High	High	High	High	High
Capacitive Sensors	High	Average	Low	High	High

TABLE 6.7   Performance of Optical and Magnetic Firmness Approaches

	Performance Metric			
	Durability	Dynamic Range	Resolution	Stability
Magnetic Instruments (NMR/MRI)	High	High	High	High
Optical (Light Scattering)	Low	High	High	High
Optical (Waveguide Deformation)	Low	Moderate	High	High

In contrast to the traditional penetrometer and other mechanical instrument–based approaches to measuring fruit firmness, microsensors (Table 6.6) can offer greater accuracy and resolution as well as greater repeatability and stability. In particular, microsensors that use piezoelectric and capacitive elements have been commercialized for a wide range of applications, and the sensor technologies are mature and offer reasonable durability and excellent stability. Piezoresistive force (and firmness) sensing technologies are also mature and often durable, but may be limited in accuracy or stability by hysteresis.

Finally, both optical and magnetic approaches to sensing firmness can experience stability, resolution, and accuracy losses due to interference from ambient light (e.g., in optical scattering), variations in other internal fruit properties (NMR), and variations in user technique (e.g., in optical waveguide deformation). The performance of both light-scattering optical and magnetic techniques for determining firmness is limited by the fact that both of these techniques are designed to capture more holistic properties of the fruit and extracting firmness out of the measurement data can be difficult. Thus, while versatile, these techniques produce data whose accuracy in measuring specific parameters like firmness is vulnerable to interference from other measurands. In addition, optical techniques that choose to use glass fibers for greater accuracy can be fragile to the detriment of their durability.

Versatility

The versatility of a sensing technology for IOT solutions refers to how well the sensor or sensing system can be adjusted to different problems, whether pre- or postharvest. Some firmness measurement technologies are implemented on instrument platforms that can be configured before delivery to serve multiple functions. Other technologies, while implemented in a fixed configuration, produce versatile data that can be processed to serve other functions in addition to or in place of sensing firmness.

In the latter category are the piezoresistive and capacitive sensors that can combine sensing force and contact area to determine firmness. While these sensors are not established technologies in agricultural applications, they have the potential to fill a niche among crops that are traditionally picked by hand and are likely to continue to be harvested by hand into the near future. Integrated into gloves, these sensors may be used to identify localized deficits in firmness (bruising), assess damage incurred by chilling temperatures, count fruit in real time, and assess quality postharvest prior to transport. When designed properly, these smart glove systems can be readily converted from ordinary picking operations to monitoring tomatoes in times of crisis.

Aside from wearable sensors and sensor arrays, several sensing instruments also offer versatility that is embedded in their general purpose design. Optical spectrometers, for instance, can be configured to sense both near-infrared and visible light, and when implemented with corresponding light sources in these wavelength bands, can be used to sense reflected light (color) as well as scattered light (firmness and other internal properties). Although best suited to research rather than individual farms and growers, magnetic NMR by its very nature gathers data that can be used to understand a broad range of internal properties of fruit, including tomatoes. Because they are designed to be more general purpose, optical, NMR, and related instruments (whether benchtop or handheld), are inherently more versatile than more customized technologies, such as penetrometers.

Of all the sensing technologies discussed in this chapter, those that are the least versatile for an IOT implementation are the traditional methods that rely on mechanical interactions with the fruit to determine firmness. Mechanical and acoustic methods that require a customized input stimulus or probe are least likely to be affected by other properties of the fruit (by design) and, for this very reason, are not versatile and not well suited to IOT-scale sensing.

CONCLUSIONS

At one level or another, all of the sensing technologies considered in this chapter are compatible with sensing firmness both pre- and postharvest. In terms of accuracy, the mechanical techniques (e.g., penetrometers, piezoresistive, piezoelectric, capacitive) have an inherent advantage because they directly measure the property of interest (i.e., pressure or force). Because they are mechanical in nature, these techniques require some level of deformation of the tomato under test, which by its very nature makes the technique vulnerable to bruising, damaging, or even destroying the fruit. However, advances in microsensor technologies can limit such damage to a tolerable level, creating only microdeformations that are negligible to the consumer and the assessment of overall fruit quality. Furthermore, microsensor technologies, especially those that are easy to integrate with integrated circuit fabrication processes, have the added advantage of smart design, where onboard electronics and data transfer to the cloud or other devices can increase both the autonomy by which firmness data are collected as well as the sheer quantity (density) of total data.

Optical, magnetic, and related techniques have the advantage of measuring the tomato with limited or no contact with the fruit itself. But, such measurements are prone to capturing more than firmness, which may or may not be a desired attribute of the sensing system. For example, the scattering of light inside a tomato measured using optical techniques is a function of a wide variety of parameters, including firmness, mealiness, total dissolved solids, Brix content, and internal defects from which firmness is inferred rather than measured directly (Butz et al. 2005).

In consideration of these limitations and benefits, those sensors that are amenable to seamless integration with everyday activities that make contact with tomatoes seem to be the most promising. Accelerometers, force, pressure, and contact sensing using piezoresistive and capacitive technologies have already been integrated into wearable packages for monitoring motion, posture, and expenditure of energy as well as identifying falls and assessing balance (Yang and Hsu 2010). Piezoresistive materials have also been woven directly into fabric to facilitate almost endless possibilities for wearable force, pressure, and strain sensing (Huang et al. 2008). Thus, cost-effective integration of sensor arrays into the fingertips of gloves is certainly within the realm of possibility for monitoring firmness during the extensive manual handling of tomatoes that characterizes many farms that grow tomatoes for fresh consumption. Wearable firmness sensors can readily provide the dense quantities of data at individual or experimental farm levels required to understand spatial and temporal variations in firmness throughout the preharvest tomato ripening process. Postharvest, these microsensor technologies are also small and flexible enough to be integrated into postharvest containers and handlers (e.g., conveyor belts and storage bins). Unlike preharvest, where these microsensors may be the only technologies evaluating firmness, in a postharvest setting, these microsensors can operate alongside other technologies, including optical, acoustic, or force-driven approaches, to provide rich information regarding the health and quality of fruit during handling and processing for transport and distribution.

CHAPTER 7

Sensing pH for the Perfect Tomato

pH STANDS FOR POWER of hydrogen, and the pH number represents the concentration of hydrogen ions in solution. It is measured on a scale between 1 and 14, with lower numbers below 6 representing acids and numbers above 9 representing strong bases. In the world of tomato farming, pH is relevant both in the soil, as tomato plants tend to prefer acidic soils between 5.5 and 7.0 (Penas and Lindgren 1990), and in the tomato itself, where pH can vary between 3.5 and 4.9 depending on cultivar, ripening stage, and end use (Clemson University Extension, 2018). Tomato pH is important because it contributes to the safety of processed tomato products and to the overall flavor of all kinds of tomatoes.

Processing tomatoes are so named because they are grown to support production of processed products such as canned diced or stewed tomatoes, catsup, and tomato juice. In these products, tomato pH must remain low enough to keep the product safe but also high enough to maintain flavor. Tomatoes are not considered a low-acid food, where "low-acid" is considered to be any food with a pH of more than 4.6. For this reason, they do not require major thermal treatments to destroy the microorganisms that can compromise food safety. Nevertheless, pH remains a food safety concern for processed tomato products with long shelf lives. In California, which produces over 90% of the U.S. processing tomato crop, most industrial processors require a pH between 4.2 and 4.3 for incoming processing tomatoes (Anthon et al. 2011). A maximum safe level of 4.4 with an optimal target pH of 4.2–4.3 has been suggested to prevent *Clostridium botulinum* (botulism) spores from growing and producing toxin (Monti 1979). Although food safety concerns due to elevated pH levels can be addressed by adding citric acid to tomato products prior to packaging, it is more cost effective to harvest tomatoes at the proper pH level than to add acid afterward.

Tomatoes grown for fresh consumption are not as vulnerable to food poisoning concerns. However, pH is still important, because chemically, pH is to flavor as color is to perceived quality with respect to the tomato. Low pH has been significantly associated with perceptions of sourness as well as astringent and metallic flavors in fresh tomatoes. The more acidic the tomato, the worse the perceived flavor as measured by multiple sensory descriptors (Tandon et al. 2003). Fortunately, pH increases as a tomato ripens. Increases in pH from the pink stage where 30%–60% of the tomato has turned pink or red in color (USDA 2005) to fully ripened can be as high as 0.4 pH points (Anthon et al. 2011). But, as tomatoes ripen more fully, they also become much more vulnerable to damage, bruising, over-ripening, and decay in transport. Therefore, determining the best time to harvest tomatoes for fresh consumption must not only take flavor into account but also the impact of handling, storage, and transport on how the tomato looks once it reaches the consumer.

To support growing and delivering the perfect tomato to the consumer, pH sensors should measure pH between 3.5 and 5.0 and have a resolution of 0.01 pH units in order to adequately track year-on-year changes in harvesting pH (Anthon et al. 2011). These sensors should also be sufficiently stable to measure tomato pH over a 60-day ripening process from mature green to pink to over-ripe, as well as during subsequent storage, handling, transport, distribution, and sale.

Except in isolated cases and in some research efforts, monitoring tomato pH has been of little interest to the production of the perfect tomato. However, understanding how pH changes over time and influences perceptions of flavor can improve harvesting strategy, increase profits, and enable more optimal selections of tomato varieties at the regional or individual farm level. Since pH sensors come in many different shapes and sizes, finding the right sensor technology is a matter of gaining a clear understanding of both the requirements and the constraints for monitoring pH during different stages in tomato production.

The simplest pH sensors use litmus or pH indicators to generate color representations of pH at coarse resolutions (on the order of 1 pH unit or greater). The most common instruments for measuring pH use an electrochemical approach with a glass electrode that includes a stable reference solution to ensure high-resolution, accurate, and consistent measurements. A wide range of other pH sensors are also available commercially. They are based on an equally broad range of sensing strategies and have attributes that are not available in glass electrodes, pH indicators, or other traditional pH sensors. And, many are smaller than glass electrodes while providing comparable accuracy and resolution, thus making them more attractive for integration into the IOT.

TRADITIONAL APPROACHES TO pH MEASUREMENT

Traditional pH sensors include litmus indicators or pH strips that indicate pH with color and glass electrodes that indicate pH as an output voltage.

Litmus indicators are among the oldest and most inexpensive pH sensors. A litmus indicator is a liquid mixture extracted from lichens that turns red when exposed to an acid and blue when exposed to a base. pH indicators that behave similarly to litmus indicators can vary from household liquids like the juice of cooked red cabbage, which is useful for at-home or school science experiments, to more consistent, commercially available litmus indicators like methyl red, which has three color ranges for highly acidic (red), moderate (orange), and highly basic (yellow) solutions. Since litmus was introduced as a way to roughly differentiate liquids by acid, base, or neutral characteristic, a wide range of pH indicators and dyes have been synthesized to provide a coarse measure of pH. Indicators that are particularly relevant to monitoring the tomato maturation and ripening process include methyl red, methyl orange, and bromocresol green, which offer color transitions within a pH unit of 5.1, 3.7, and 4.7, respectively (Pennsylvania State University n.d.). Naturally occurring pH indicators that offer color transitions in the pH ranges that correspond to tomato ripening processes include red cabbage, turnip skin, and the blue dye found in Asian pigeon wings (Khankaew et al. n.d.).

pH strips also provide rough measurements of pH and do so across a much broader dynamic range than pH indicators alone. Strips combine multiple indicators into a single color-based sensor. Once exposed to a solution of interest, the color of a pH strip is either manually compared to a color chart with the naked eye to determine pH or assessed by a specialized instrument called a colorimeter to determine pH from strip color. Like litmus and other pH indicators, pH strips provide a resolution of about 0.5 pH units, albeit across a wider dynamic range than single indicators. This level of resolution may be useful for distinguishing a green tomato from a fully ripe one, but beyond that, a pH sensor must deliver greater resolution to be useful for assessing tomato quality throughout growing, harvest, and postharvest handling.

Glass electrodes offer much higher-resolution and more sophisticated measurements than pH indicators and strips. The glass electrode (Figure 7.1a) uses an electrochemical approach to converting pH to an output voltage. Hydrogen ions in solution interact with a solid material, called a working electrode, to either extract or impart electrons to the electrode. The number of electrons that are exchanged between the solution and the working electrode depends on the concentration of hydrogen ions in the solution. Hydrogen ion concentration ($[H^+]$) in dilute solutions is related to pH as:

$$pH = -\log_{10}([H^+]) \tag{7.1}$$

In order to measure the exchange of electrons between hydrogen ions in solution and the working electrode, a second electrode, called a reference electrode, is used. The reference electrode material and surrounding solution are chosen for consistent and reliable electron exchange between the two. This exchange produces a known and stable potential with which to measure the potential on the working electrode. A typical glass electrode system hosts an Ag/AgCl (silver/silver chloride) reference electrode in a constant pH solution in order to establish the reference potential and another glass electrode coated with a hydrated gel to serve as the working electrode when exposed to a solution of unknown pH. The resulting voltage between working and reference electrodes follows the familiar Nernst equation:

$$V = V_o + \frac{RT}{nF} \ln \frac{[H]_{reference}}{[H]_{measure}} = V_o + \frac{RT}{nF} 2.3 \log_{10} \frac{[H]_{reference}}{[H]_{measure}} \tag{7.2}$$

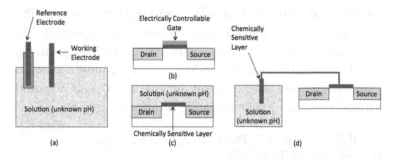

FIGURE 7.1   Electrochemical approaches to measuring pH. Traditional approaches to accurately measuring pH use a glass electrode–based system that relies on (a) a reference/working electrode pair that measures an unknown concentration of ions in solution. Miniaturized approaches to measuring pH rely on the (b) standard electronic field effect transistor structure with the gate removed to create (c) a chemically sensitive ISFET. Electronics and chemistry can be separated in the ISFET by using (d) an extended gate structure.

Converting this expression to pH gives:

$$V = V_o + \frac{2.3RT}{nF}(\text{pH}_{measure} - \text{pH}_{reference}) \tag{7.3}$$

where V_o, R, n, and F are constants for a given type of electrode or pH sensor; T is temperature in degrees Kelvin; $[H]_{reference}$ and $\text{pH}_{reference}$ are the concentration of hydrogen ions and pH in the reference solution; and $[H]_{measure}$ and $\text{pH}_{measure}$ are the concentration of hydrogen ions and pH in the sample solution. Nernstian measurements are stable, consistent, and reliable. They also do not suffer penalties of scale, meaning that a smaller sensor based on Nernstian behavior produces as accurate an output as a larger sensor.

Glass electrode pH measurement systems based on the Nernstian response to pH are commercially available from multiple manufacturers, including Hach, Fisher Scientific, Cole-Parmer, Hanna Instruments, and Mettler Toledo, and cost between $100 and $500. These instruments have been used to successfully monitor tomato maturation (Hanna 2015). However, they require that a tomato be sampled before measurement, thus causing damage to the fruit that makes it no longer marketable. The glass electrode instrument is also relatively large, expensive, and fragile, and requires frequent recalibration. Nevertheless, among all pH sensors, the glass electrode has been the most successful commercially and should always be considered as a potential solution to needs for pH monitoring within agriculture.

OTHER ELECTROCHEMICAL APPROACHES TO pH MEASUREMENT

Demand for smaller, cheaper pH sensors has stimulated extensive research in miniaturized electrochemical sensor designs (including the glass electrode) for pH and other analytes (Janata 2003). The most common miniaturized version of the glass electrode electrochemical sensor is the ion-sensitive field effect transistor (ISFET) shown schematically in Figure 7.1c. The ISFET works by leveraging the electrical behavior of a field effect transistor (FET, Figure 7.1b) to capture the exchange of electrons between the surrounding solution and the surface of the ISFET. In a conventional FET, the top layer of the transistor is a metal (or other conductive) layer that controls the current flowing in the underlying transistor (between the drain and the source). For example, in a conventional n-type metal oxide semiconductor field effect transistor (MOSFET), a positive voltage applied to the top layer or gate attracts electrons from the substrate to the gate, where they form a conductive channel for more electrons to travel through the transistor (Figure 7.1b). The surface layer of an ISFET can serve the same purpose as the positive gate voltage in the MOSFET. By donating electrons to the surrounding solution, the top layer of the ISFET produces a net positive charge that has the same overall effect on the underlying transistor current as the applied positive voltage in a conventional n-type MOSFET. In a pH-sensitive ISFET, the metal or conductive gate of a MOSFET is removed and charge accumulates on the insulator layer of the device, thereby controlling the current in the transistor. The insulator layer is chosen so that it either donates (n-type FET) or accepts (p-type FET) electrons from the surrounding solution in a manner that follows the Nernst equation (7.3). Silicon dioxide and silicon nitride are two common choices for the ISFET insulator layer. These two materials are not only sensitive to pH but are widely used as the insulator layer in conventional transistors. As a result, these materials are highly compatible with established integrated circuit fabrication processes and the reductions in cost that go with those processes.

Despite being introduced to pH sensing in the 1970s, the ISFET was not commercialized until the 1990s. In part, commercialization was delayed because ISFETs are inherently vulnerable to drift and sensitive to changes in both light and temperature, thus limiting their accuracy and resolution. Furthermore, like the glass electrode, an ISFET still requires a reference electrode to produce a stable voltage to which the potential on the active or working ISFET can be compared. Similar to the traditional glass electrode system, a silver or mercury chloride reference electrode is preferred for stability, but these electrodes are often bulky and fragile, thus limiting opportunities to make a truly miniaturized ISFET. For food quality monitoring in particular, possible leakage from a mercury chloride reference also poses a health risk. But, in spite of these limitations, some pH ISFETs have emerged in commercial markets such as those produced by Emerson, ThermORION, Honeywell, and Sentron. Unfortunately, however, the cost of these sensors can approach that of conventional glass electrode systems due to the need for a bulky reference electrode. To address this limitation, conventional reference electrodes can be avoided by either taking a different approach to sensor design or by miniaturizing a conventional silver/silver chloride electrode. The alternative reference FET (REFET) works by making two ISFET structures where the first structure is pH sensitive and the second is not (Chudy et al. 1999). The second device operates similarly to a reference electrode but has a shorter lifetime than conventional reference electrodes, which has limited its commercial appeal. Miniaturizing the traditional reference electrode design has been more successful, and MicroSens makes an integrated ISFET-based pH sensor using such a miniaturized electrode.

A variety of miniaturized electrochemical sensors based on the ISFET principle have been developed to overcome limitations in ISFET systems. The extended gate FET (EGFET, Figure 7.1d) separates the electrical and chemical components of the ISFET system to preserve the lifetime of the electronics, promote stability, and limit drift. While sensing pH, only the extended pH-sensitive gate and reference electrode are immersed in solution, while the remainder of the FET structure and interface electrodes remain dry and in the dark. The EGFET is far more stable than the ISFET because the FET part of the structure is not exposed to liquid or light and temperature variations that cause drift and instability in traditional ISFETs (Qin et al. 2015). The EGFET approach also allows the electrochemically sensitive portion of the sensor to be fabricated with custom and optimized designs while corresponding transistors and interface circuits can still be made using standard microfabrication techniques. The net result is a more stable and lower-cost sensor than is possible with traditional ISFET designs.

For higher resolution, dual-gate ISFETs (Spijkman et al. 2011) contain two gates rather than one and by manipulation of the ratio of top to bottom gate capacitance have increased the Nernstian limited sensitivity of 59 mV/pH unit by one or more orders of magnitude. Other electrochemical sensors are also available that measure a current proportional to pH in solution, but the performance of these amperometric devices decreases dramatically with miniaturization, thus making them better suited to larger instrument designs rather than miniaturized sensors. In contrast, all potentiometric sensors, including the ISFET, REFET, and EGFET, are largely immune to scaling penalties; their performance does not decline with diminishing size.

ELECTRICAL APPROACHES TO pH MEASUREMENT

As an alternative to electrochemical conversion, some pH sensors can act by converting pH changes directly to a change in a passive electrical parameter such as resistance. Converting pH to resistance (or conductance) is particularly attractive because subsequent interface circuitry and signal transmission can be compact and low power. These types of sensors can also be made using standard microfabrication processes, which can lower their cost tremendously compared to the specialized fabrication techniques required for many optical, electrochemical, and other pH sensors. Multiple materials and structures respond to pH in a way that can be measured via conductance or resistance. Two established classes of materials that can be used to measure pH in this way are conducting polymers and metal oxides.

Conducting polymers make good candidates for sensing of pH via resistance or conductance. These conducting polymers are well suited to detect molecules in gases because they are highly sensitive, respond quickly, can operate at ambient temperature, and have mechanical properties suitable for easy fabrication (Bai and

Shi 2007). To a lesser extent, in a liquid environment, conducting polymers contain certain molecules that can respond to changes in pH. In these polymers, the surrounding liquid solution extracts electrons from the aromatic rings in the polymer, thereby causing an increase in the apparent doping of the p-type polymer and a corresponding increase in conductance. pH-sensitive conducting polymers include polyaniline or PANI (Arshak et al. 2007), which is particularly effective at the low pH ranges that are relevant to tomato maturation, ripening, and postharvest handling. Stability issues associated with these conducting films can be addressed through the addition of other polymers like polypyrrole (Gill et al. 2008a).

Like conducting polymers, metal oxides are best known for gas sensing, but some can sense pH in liquids too. For example, ruthenium oxide deposited on an interdigitated silver electrode experiences a decrease in resistance with increasing pH within the range of 3–11 pH units (Hristoforou and Vlachos 2014). When a DC bias is applied, titanium monoxide is also sensitive to pH in solution, but poor repeatability makes this material suitable only for rough or single-use measurements (Gill et al. 2008b). At low frequencies, different types of tin oxide films also show an increase in conductance with pH in the range of 2–11 pH units (Gill et al. 2008b). pH-sensitive metal oxides are especially attractive for their superior chemical stability compared to many other pH-sensing technologies.

Electrochemical pH sensors face serious challenges in miniaturization, not the least of which is in finding a stable, suitable reference electrode. In contrast, electrical pH sensors based on conductivity or resistivity are highly scalable and well suited to a small, low-cost, low-power footprint, which can make them a more attractive for sensors in the IOT.

ELECTROMECHANICAL APPROACHES TO pH MEASUREMENT

In an electromechanical sensor, pH is converted to a change in one or more mechanical properties of the sensor (e.g., bending, mass, shape, elasticity). These mechanical changes are intermediates that are subsequently converted to an electrical output. One of the most popular designs for an electromechanical sensor is the micro- or nanoscale cantilever beam. The beam is coated with a material that responds to pH by changing size, shape, elasticity, mass, or some other mechanical property. The change in pH-sensitive material deforms the beam and the deformation is measured using piezoresitive, capacitive, or similar techniques to identify pH.

For example, hydrogels are a class of materials that undergo significant conformal changes in shape and size in response to changes in pH and, when adhered to a cantilever beam, cause appreciable bending as a result. A hydrogel is a polymer that contains molecular chains that are cross-linked into a three-dimensional matrix. This dense web of interconnections or cross-linking makes hydrogels insoluble in most solvents. But, this cross-linked web also causes the hydrogel to swell in the presence of water. In one phase, the hydrogel is at its most hydrophobic (i.e., repelled from water) and shrinks. In a second phase, the hydrogel tries to mix with the solution in which it is immersed, thus causing it to expand or swell at maximum hydrophilicity (i.e., attraction to water). Some hydrogels are stimuli responsive and transition from one phase (hydrophobic) to the other (hydrophilic) based on pH, temperature, or other properties of the solution in which they are immersed (Richter et al. 2008). Hydrogels can experience large changes in shape up to 100-fold in response to small changes in pH. Acidic hydrogels are particularly well suited to tomato monitoring applications because they experience these large changes in size and shape in acidic liquids. Used in conjunction with cantilever beam technology, hydrogel sensors have been demonstrated with resolutions on the order of 10^{-5} pH units and movements ranging from 1–20 microns per pH units (Richter et al. 2008).

High sensitivity limits dynamic range, however, as tiny cantilever beams can only bend so far. The large conformational change that stimuli-responsive hydrogels experience in response to pH changes may be too much for cantilever platforms, because these beams typically bend no more than a micron in either direction. To solve this problem, other pH-sensitive materials such as silicon nitride and silicon oxide can be deposited on microcantilevers. Ionization of species on the surface of these materials causes beam bending which, like hydrogel-coated beams, can be measured as a change in resistance by the integration of piezoresistive sensing elements onto the beam. As the degree of ionization changes with pH, the microcantilever beams can bend in response, typically across a smaller dynamic range than hydrogels, thus making for a more practical pH microsensor (Boisen and Thundat 2009). Although other structures such as diaphragms are available to facilitate electrochemical sensing, the microcantilever beam approach is both compatible with conventional microfabrication processes and highly sensitive.

Other sensitive electromechanical approaches to pH sensing have used the quartz crystal microbalance (QCM). The QCM is piezoelectric and converts a time-varying mechanical change to a time-varying voltage, rather than to a resistance as is the case with piezoresistive elements. When a QCM is coated with hydrogel, the frequency at which the time varying signal produces the greatest change (i.e., the resonant frequency) changes with both increasing mass and volume of a hydrogel, thus making a highly sensitive pH sensor (Richter et al. 2008) without the dynamic range limitations of a cantilever beam sensor.

While interesting and offering high sensitivity and high resolution for pH measurement, electromechanical approaches, because of their increased complexity, tend to be reserved for analytes (chemicals) that have few if any lower-cost alternatives. Thus, the wide range of alternative approaches for pH sensing limit how attractive electromechanical approaches are for monitoring pH in tomatoes.

OPTICAL APPROACHES TO pH MEASUREMENT

In addition to electrochemical, electrical, and electromechanical sensors, optical approaches can also be effective for measuring pH. In an optical system, pH changes are first converted to changes in a light signal, and then into a corresponding change in an electrical parameter like voltage.

Most optical pH sensors involve a pH-sensitive dye or other material that is immobilized in a solid material or matrix. This matrix is designed so that that the protons or hydronium ions that are indicative of the pH in the surrounding solution permeate throughout the matrix. As these ions make contact with the pH-sensitive dye, they alter its optical properties. In many dyes used for optical pH sensors, the amount of light absorbed or the wavelength at which it is absorbed changes with pH. Ocean Optics has commercialized optical pH sensors based on this approach using sol-gel and metallized materials whose absorption properties allow for measuring pH between 5.0 and 9.0.

Other optical sensors rely on pH-sensitive layers whose fluorescent behavior changes with pH. Fluorescent pH dyes work by absorbing light across a relatively wide range of colors or wavelengths and re-emitting that light at a different and distinct color. In many materials designed for this purpose, fluorescence intensity (i.e., the amount of emitted light) changes with pH. However, some fluorescent dyes exhibit changes in color, lifetime, or phase as pH changes in solution. Resolutions on the order of 0.1 pH units are possible, and detectors coupled to fluorescent dye systems must only sense light within a very narrow bandwidth of wavelengths, thus providing greater selectivity, sensitivity, and robustness than other forms of optical pH sensing. Fluorescing dyes that detect pH in the 3.5 and 4.9 range relevant to tomatoes have been demonstrated (Wencel et al. 2014). This approach has also been commercialized. For example, CellPhase makes a pH sensor based on an organic polymer that uses the ratio between two different fluorescent behaviors to determine pH. PreSens also uses an organic polymer in its commercial products, but instead relies on changes in the lifetime of the fluorescent light to determine pH (Wencel et al. 2014).

The refractive index (i.e., the ratio of the speed of light in a vacuum to its speed in another medium) is another optical property of liquids that can be leveraged to determine pH. Using refractive index (RI) as an indicator of pH has some advantages over using color indicators. While color can be measured with the naked eye, as is the case with conventional pH indicators and strips, the eye can be inconsistent, unreliable, and imprecise. More precise measurements of color require a colorimeter (absorption of light) or a spectrophotometer (transmittance or reflectance of light) to identify characteristics of the indicator at specific wavelengths. Such instruments are expensive and of limited use for pH sensing. To avoid the need for a spectrophotometer or colorimeter, pH may be measured optically using a refractometer instead. RI changes predictably with pH, and methods ranging from very inexpensive ($10), lower-resolution instruments to very high-resolution instruments based on such phenomena as surface plasmon resonance are commercially available to reliably measure refractive index. The primary drawback of using RI as a pH indicator, however, is that RI changes with several parameters, including sugar content and total suspended solids, that vary concurrently with pH during tomato maturation, ripening, and degradation.

RI changes can indicate many things and thus make for a poorly selective pH sensor. However, when used in conjunction with materials that respond to pH and only pH, RI measurements become much more attractive. Instruments based on surface plasmon resonance (SPR) offer both the opportunity to exploit the pH preferences of certain materials and measure the resulting RI changes at very high resolutions. For example, RI changes in hydrogels can be quite significant during phase transitions and are easily picked up using SPR across a wide dynamic range from 1–12 pH units (Zhao et al. 2018). Unfortunately, SPR has been commercialized primarily as a benchtop laboratory instrument rather than as a portable, low-cost sensor, thus limiting its application to tomato growing and production.

Regardless of whether absorption, reflection, fluorescence, refractive index, or some other optical property changes with pH in a material, the signal that results from detecting these changes is also optical. In most cases, this requires another sensor to convert the light signal into an electrical one. Typically, this involves transferring light from where it is collected to where it can be measured with the goal of losing as little light as possible in the process. Low-loss light transfer can be done using either optical fibers or planar waveguides. The optical fiber has been around for decades, is available in glass and in a less fragile but less efficient (i.e., more lossy) plastic counterpart, and is vulnerable to misalignment, which produces both loss of signal and loss of accuracy. Planar waveguides do the same thing as optical fibers by guiding light from its point of origin to a light detector or sensor but are more robust, less likely to become misaligned, and better suited for miniaturization and integration with circuits and sensors. Whether by fiber or by planar waveguide, once the optical signal reaches its destination, it can be converted to an electrical signal using charge coupled devices, photodiodes, phototransistors, or other light detectors. A wide range of light detectors are available commercially, are well characterized, and measure light accurately and repeatably.

Optical approaches to pH sensing are attractive because they are less vulnerable to interference than other techniques and can capture pH in the acidic range of interest (3.5–4.9) for monitoring pH in tomatoes at high resolutions. Furthermore, unlike electrochemical sensors, which require that the tomato sample come into contact with sensor materials that may not be food-grade safe, optical sensors can use naturally derived (Khankaew et al. n.d.) and synthetic dyes, which are safe for contact with food. However, like electromechanical approaches, optical methods involve multiple stages of conversion before an electrical output is produced and, for this reason, tend to be bulkier, more complex, and more expensive than other sensor approaches.

SEEKING THE pH OF THE PERFECT TOMATO IN THE INTERNET OF THINGS

Monitoring pH in a tomato is important at a different scale and scope than monitoring color and firmness. For example, color and firmness of tomatoes grown for fresh consumption are relevant at the individual fruit level from farm to table. Preharvest, color and firmness determine whether a tomato is ready to be picked and can indicate bruising and other damage. The color and firmness corresponding to optimal harvest times are dependent on how long and how far the tomato has to go before it lands on the consumer's table. In an ideal world, the color and firmness of each tomato would be measured multiple times before it reaches the consumer. For pH, however, this level of monitoring would be excessive. A more suitable approach would be to monitor pH by tomato variety, by regional weather and climate patterns, or by consumer perceptions of flavor and taste. Knowing the pH of lots of individual tomatoes is irrelevant, but understanding pH trends at the batch, crop, variety, regional, or similar level is very relevant. With this in mind, the suitability of various pH sensor technologies for monitoring tomatoes is summarized in Table 7.1.

Accessibility (Table 7.2)

It would be difficult to surpass traditional pH indicators and litmus strips in terms of accessibility. Performance notwithstanding, this centuries-old approach to detecting pH is commercially available in a wide range of products, consumes no power, and is low cost, compact, and easy to use. But, when other considerations such as performance, compatibility, and versatility come into play, other pH sensors may be of more interest. Glass electrodes (electrochemical sensors) are widely available commercially and are reasonably affordable at less than $500 but also tend to be bulky and power hungry and require periodic recalibration and maintenance, which limits their ease of use over the growing season. Optical technologies that rely on fluorescence or absorption are also configured as instruments rather than as sensors and tend to cost as much or more than glass electrode–based instruments, which limits their accessibility to individual farmers or other small tomato production operations. Electrical and electromechanical technologies have been demonstrated

TABLE 7.1 Suitability of pH Sensors for IOT Solutions

| | IOT Characteristic | | | | | |
	Overall	Accessibility	Compatibility	Density	Performance	Versatility[a]
Litmus Indicators	Low	High	Low	Low	Low	Low
pH Color Strips	Low	High	Low	Average	Low	Low
Electrochemical (Glass Electrode)	Average	Average	Low	High	High	Low
Electrochemical (ISFET)	Average	Average	Average	High	Average	Low
Resistance Sensors (Polymer)	High	High	High	High	Average	Low
Resistance Sensors (Metal Oxide)	Average	High	High	High	Average	Low
Electromechanical (Hydrogel)	High	Average	High	High	Average	Low
Electromechanical (Other)	High	Average	High	High	High	Low
Optical Sensors (pH Indicators)	Average	Low	Average	High	Average	Low
Optical Sensors (Fluorescence)	Low	Low	Low	High	High	Low

[a] Although in the conventional sense, pH sensors are designed with low versatility (i.e., they sense pH and only pH), they can be versatile in other ways both pre- and postharvest.

TABLE 7.2   Accessibility of pH Sensor Technologies

	Accessibility Metric				
	Affordability	Availability	Battery Life	Compactness	Ease of Use
Litmus Indicators	High	High	High	High	High
pH Color Strips	High	High	High	High	High
Electrochemical (Glass Electrode)	Average	High	Average	Average	Average
Electrochemical (ISFET)	Average	Low	High	High	High
Resistance Sensors (Polymer)	High	Low	High	High	High
Resistance Sensors (Metal Oxide)	High	Low	High	High	High
Electromechanical (Hydrogel)	High	Low	Average	High	High
Electromechanical (Other)	High	Low	Average	High	High
Optical Sensors (pH Indicators)	Average	Average	Low	Average	Average
Optical Sensors (Fluorescence)	Average	Average	Low	Low	Low

as compact, battery-compatible sensors but have very limited commercial availability at this time.

While accessibility of pH sensors for monitoring tomato growing and other operations may seem limited, it is important to remember that not every farm or operation need have a pH sensor(s) or instrument on hand at all times. Rather, opportunities for pH sensors to increase the success of tomato growing and production lie in collecting and sharing data that characterize batches or varieties of tomatoes rather than monitoring individual fruit.

Compatibility (Table 7.3)

The compatibility of pH sensors with the IOT is ill suited to conventional notions of autonomy where data are collected with

TABLE 7.3   Compatibility of pH Sensors with Sensor Platforms

	Platform					
	Preharvest			Postharvest		
	Benchtop	Handheld	Wearable	Storage	Conveyer	Package
Litmus Indicators		✓				
pH Color Strips		✓		✓		✓
Electrochemical (Glass Electrode)	✓	✓			✓	
Electrochemical Microsensors (e.g., ISFET)	✓	✓	✓	✓	✓	✓
Resistance-based Sensors (Polymer)	✓	✓	✓	✓	✓	✓
Resistance-based Sensors (Metal Oxide)	✓	✓	✓	✓	✓	✓
Electromechanical Sensors (Hydrogel)	✓	✓	✓	✓	✓	✓
Electromechanical Sensors (Other)	✓	✓	✓	✓	✓	✓
Optical Sensors (pH Indicators)	✓	✓			✓	
Optical Sensors (Fluorescence)	✓	✓			✓	

little to no human intervention and seamlessly transferred to other devices or the cloud for interpretation. The day when a talented ground robot can roam rows of crops and insert the smallest of probes into one tomato after another to monitor pH without risking damage to the fruit and without a human being overseeing the process is likely far into the future. The same is true for using unmanned aerial vehicles to monitor pH.

Instead, in the short term, measuring pH will likely continue to require that a tomato be picked, sampled, and processed (i.e., juice extracted) prior to measurement. Making this process more accessible to a wider range of constituents, however, can have an impact on the autonomy of data sharing through IOT pathways.

For example, in a world where low-cost, digitized pH sensors can wirelessly connect to smart phones, a meaningful farm-to-table link can become a reality. Individual farmers might monitor pH through maturity and ripening, while subsequently consumers evaluate taste and flavor at the dinner table and enter their impressions in a smart phone app that relays such information back to the farmer. The link between what the farmer provides and what the consumer receives, particularly as it relates to taste, can make a world of difference in better meeting consumer demand and expectations. Similar data links between tomatoes grown for tomato products and processing facilities can support more autonomy in how and when tomatoes are harvested, transported, and queued for transport into large-volume tomato processing facilities. Thus, integrating pH sensors into an IOT approach to growing, harvesting, and marketing tomatoes takes a different approach than other sensors like those that collect firmness and color data, but can still leverage and benefit from the massive interconnections and communication among constituents that the IOT can provide.

Unfortunately, pH sensors, by their necessity for direct sampling and contact with the juice inside of a tomato, have limited compatibility with existing infrastructure on farms and postharvest handling and transport operations. Because pH sensing almost always requires direct contact with the fruit, aerial platforms, both manned and unmanned, are out of the question. The integration of pH sensors onto ground vehicles, whether manned tractor or unmanned robot, is also unlikely.

Traditional instruments used for measuring pH (i.e., the glass electrode) are commercially available as benchtop and handheld devices, but, due to the fragility of the glass probe, may be best suited to benchtop sampling and testing that occurs off-field or away from a production line. Litmus indicators and pH strips also have limited compatibility with monitoring tomatoes, because measuring and recording data requires a great deal of human oversight.

In principle, miniaturized electrochemical sensors, electromechanical sensors, and resistance sensors bridge the gap between traditional glass electrodes and pH indicators in terms of their compatibility with existing infrastructure and capacity for collecting data. In particular, resistance sensors offer the same single-use, ultra-low-cost advantages of pH indicators while producing a more accurate and electrical output that is compatible with a wide range of platforms. If coupled with miniaturized extraction mechanisms (e.g., microneedles) that can siphon a small volume of juice from a tomato, resistance-based sensors are compatible with a full range of platforms, including handheld monitors and on-field wearables, for preharvest pH monitoring as well as storage and packaging containers for postharvest needs. Optical techniques that use microneedles to sample pH are also, in principle, compatible with most sensor platforms available to those involved in tomato growing and handling, although their size and cost limit their use in wearable devices as well as in storage bins, pallets, crates, and packing.

Density

pH sensors are not well suited to generate large quantities of data at the raw data or sensor level. The collection of data from these sensors must be approached in a different way in order to achieve a worthwhile advantage over present practices.

Leveraging IOT approaches with firmness (Chapter 6) and color (Chapter 5) measurements makes the most sense when these properties can be monitored for individual tomatoes and a large percentage of a tomato crop or batch. With pH, however, it makes little sense to monitor the pH of individual tomatoes within a batch or crop, in part because color sensors can do so more efficiently and cost effectively. pH is also a less accurate measure of ripeness and does not at all indicate internal damage or structural integrity like firmness sensors can. What then, does measuring pH at the dense, highly interconnected scale of the IOT offer? To answer this question, it is necessary to examine the two primary needs for sensing the pH of a tomato.

The harvesting, sorting, and transport of tomatoes grown for processed tomato products are often highly mechanized. While tomatoes are an acidic food, pH of the final product must still be maintained within a certain range to ensure food safety while maintaining acceptable taste. A critical piece of improving the economics of making processed tomato products is to maximize the use of processing facilities. Doing so often requires that tomatoes "wait" in the field until it is their turn to head into these processing facilities. But, as tomatoes wait, they continue to ripen and their pH increases. pH can increase between .01 and 0.02 pH units per day during these waiting periods, leading to large changes over harvest delays that last as long as four weeks. While titratable acidity, a companion metric to pH, also changes significantly during these wait periods and is dependent on tomato variety, other measures of interest, including Bostwick consistency values, color, lycopene content, total pectin, and water solubility, do not (Anthon et al. 2011). Significant increases in pH during wait periods can necessitate that citric acid be added to processed tomato products at the point of canning and packaging. The cost of these additives can offset any cost or efficiency gains made by maximizing the use of processing facilities. Thus, during field waiting periods, monitoring pH or titratable acidity of a statistically representative population of tomatoes is valuable to the success of the entire crop. Dense temporal data collected within a season and from season to season from a large number of farms in a particular area can be used to better adjust the flow of fruit from the fields to processing facilities and thereby reduce the detrimental impacts of delayed harvests incurred by processing bottlenecks.

pH sensors can also be used to gather data and inform harvest and processing practices on farms that grow tomatoes for fresh consumption. On these farms, pH is less relevant to food safety and more important to flavor and taste. pH can be used to complement more visible consumer quality indicators like firmness and color to deliver a better product to consumers at a greater profit to the grower. As with processing tomatoes, however, it is not necessary to monitor the pH of each individual tomato. Instead, the benefit of an IOT solution to sensing pH is both in characterizing pH changes over maturation and ripening time with high temporal resolution and correlating and analyzing that data in conjunction with soil pH, temperature, weather, and other relevant data collected and aggregated in the cloud. The end result is a dense, rich, detailed picture of how pH varies with tomato variety, growing conditions, and handling practices.

Performance (Table 7.4)

The performance of pH sensors is highly dependent on technology and varies greatly. For example, litmus indicators and pH color strips offer performance suitable only for rough estimates of pH and are poorly suited to resolutions of 0.1 or below. However, for quick, coarse estimates of pH using a disposable sensor, the litmus indicators and pH indicators can't be beat. Unfortunately, their resolution is often too low to be of much use for monitoring pH as relevant to flavor or food safety.

The performance of the traditional glass electrode remains the best choice in terms of accuracy, resolution, and dynamic range, but its durability in the field or on the production line can limit its applicability to tomato monitoring. Miniaturized versions of the glass electrode that improve the durability of the electrochemical approach, however, still suffer from frequent stability problems caused by reference electrode miniaturization and integration.

Electromechanical approaches that use a beam or other mechanical structure mated with a pH-sensitive material have demonstrated excellent performance and have the potential to compete with the glass electrode pH sensor in terms of resolution, accuracy, and stability. These sensors rely on the movement of a mechanical structure, typically measured as a change in the resonant frequency of that structure, to monitor changes in mass, elasticity, or shape of part or all of the structure. With customized coatings, the resolution and accuracy of these electromechanical pH sensors have been excellent. For lower cost, hydrogels offer an inexpensive solution at a reduced dynamic range. Metal oxide sensors, because of their excellent chemical stability, may also be

TABLE 7.4   Performance of pH Sensor Technology

| | \multicolumn{5}{c}{**Technology Characteristic**} | | | | |
	Accuracy	**Durability**	**Dynamic Range**	**Resolution**	**Stability**
Litmus Indicators	Low	Moderate	Low	0.5[a]	Low
pH Color Strips	Low	Low	High	0.5[a]	Low
Electrochemical (Glass Electrode)	High	Moderate	High	0.01[b]	High
Electrochemical (ISFET)	Moderate	High	High	0.002[b]	Moderate
Resistance Sensors (Polymer)	Low	High	Moderate	0.5[c]	Low
Resistance Sensors (Metal Oxide)	Moderate	High	Moderate	0.05[d]	High
Electromechanical (Hydrogel)	Moderate	High	Moderate	0.02[e]	Moderate
Electromechanical (Other)	Moderate	High	Moderate	0.0001[e]	Moderate
Optical Sensors (pH Indicators)	Moderate	Moderate	Moderate	0.001[e]	Low
Optical Sensors (Fluorescence)	High	Moderate	Low	0.1[f]	Moderate

[a] Adapted from Khan, M. I. et al. 2017. *Microsystem Technologies* 23 (10): 4391–4404.

[b] Adapted from Sea-Bird Scientific. 2018. "Comparing ISFET and Glass Electrode PH Sensors." 2018. https://www.seabird.com/cms-portals/seabird_com/cms/documents/casestudy-isfet-glass_0.pdf.

[c] Adapted from Gill, E. et al. 2008a. "Conductimetric PH Sensor Based on Novel Conducting Polymer Composite Thick Films." In: *2008 31st International Spring Seminar on Electronics Technology*, 478–483. Budapest, Hungary: IEEE.

[d] Adapted from Simic, M. et al. 2017. *IEEE Sensors Journal* 17 (2): 248–255.

[e] Adapted from Yuqing, M. et al. 2005. *Journal of Biochemical and Biophysical Methods* 63 (1): 1–9.

[f] Adapted from Wencel, D. et al. 2014. *Analytical Chemistry* 86 (1): 15–29.

a good choice for integration into wearable, field, storage bin, or conveyer belt sensing systems.

Optical techniques for measuring pH also have limitations in terms of their performance. For example, optical pH sensors based on dyes can have low stability, as dyes can age, photo-bleach, and change color in solution quite rapidly. Exposure to ambient sunlight or to heterogeneous and nontransparent tomato juice samples can ultimately degrade their performance dramatically. Fluorescent sensors often have limited dynamic range because their fluorescent properties are tailored to specific applications.

While no pH sensor delivers perfect performance, many still perform well enough to be used at various phases of bringing tomatoes from the fields to the table. The fact that pH sensors used for monitoring tomatoes need not be high resolution and do not require a broad dynamic range means that most of the pH sensor technologies in Table 7.4 perform at or beyond what is needed.

Versatility

Historically, chemical sensors have been researched, designed, and commercialized to sense a single parameter and to be impervious to all molecules except the target molecule. In this context, almost every one of the pH sensors considered in this chapter is not versatile at all. Litmus indicators, color strips, electrical, electrochemical, and electromechanical pH sensors of all kinds, as well as most optical sensors that rely on pH-sensitive dyes or fluorophores detect pH and little else. Refractometers are a classic example of why chemical sensors are designed with such narrow selectivity. While a change in pH does produce a change in refractive index, many other things, chemical and otherwise, can also influence refractive index. For example, in measurements of fruit, refractive index can just as well indicate a change in sugar content as a change in pH or a mixture of both. Thus, refractive index sensors are poorly suited to measuring pH in fruit, including tomatoes, *because* of their versatility.

But, the pH of a tomato is only one pH of interest prior to harvest. The versatility of pH sensors can be expanded by their application to multiple media. For example, soil pH is a critical contributor to plant health and can have a significant impact on crop yield through blossom rot and other ill effects (UGA Extension 2017), as well as the nutritional value of the harvested tomato (Dorais et al. 2008). In addition, the pH of irrigation water can compromise nutritional availability in tomato plants and threaten both crop health and yield. Irrigation water with pH less than 5.5 can hurt plants, particularly young roots and leaves; damage irrigation equipment; and reduce the efficiency of pesticides (Park et al. 2014). Highly acidic irrigation water (pH < 4) can lead to soil acidification and compromise nutritional availability over longer time scales if left untreated. Fortunately, most pH sensors discussed in this chapter can be used to measure pH in irrigation water and soil samples just as readily as in tomatoes, thus providing versatility by sensing pH in multiple media. Comprehensive measurements of pH in soil, water, and fruit throughout the growing season can also provide valuable insight into the impact of acid levels in these three media (soil, water, fruit). Data can not only help individual farmers increase yields through optimizing soil treatment and irrigation strategies, but, by data sharing through interconnected IOT channels, can support greater fruit production at the regional level and beyond.

At a functional level, pH becomes much more versatile by virtue of the fact that pH is strongly associated with multiple other properties of tomatoes, including Vitamin C content, ripening stage, color, taste, titratable acidity, and tomato variety (Anthon et al. 2011, Gutheil et al. 1980, Heflebower and Washburn 2010). Preharvest measurements of pH can be used to infer these parameters, especially when tracking pH changes throughout the ripening process and interpreting these changes in the context of tomato variety, weather and temperature patterns, and other data available from the cloud. Postharvest, pH tracking can be used to monitor food safety as well as to predict changes in flavor as tomatoes are treated or begin to over-ripen.

CONCLUSIONS

Overall, pH is a useful parameter that, while relevant to improving the value and yield of tomatoes harvested for multiple purposes, should be measured at a different level and scope than more traditional parameters like color and firmness. For correlating flavor to pH, where high accuracy and resolution are essential, benchtop or handheld instruments, such as the traditional glass electrode, can be the best choice. These correlations can be made at the research or experimental farm level for regional tomato varieties and the data shared via the IOT with individual farms as guidelines for determining optimal harvest parameters. At the individual farm level, processing tomatoes need to be monitored daily for pH changes that occur while waiting in the fields for entry into processing plants. Even in this scenario, however, it is not necessary to monitor individual tomatoes, but to understand how representative populations of tomatoes are trending from day to day in order to alleviate food safety concerns without costly additions of citric acid or other additives to adjust pH right before canning and packaging. Monitoring populations of tomatoes while they are in processing queues can also be done with conventional glass electrode instruments or, alternatively, miniaturized versions of those instruments using ISFETs or similar approaches. The low-cost and miniaturized pH sensing technologies may seem of little use in these scenarios, but are often the only options for tracking pH while tomatoes are in storage, transport, or packaging.

With pH sensing, there is no single best choice for an IOT-scale solution in the world of sensing for the perfect tomato. Instead, the best choice of sensor must take into account where pH is being sensed, for what purpose it is needed, and how extensively data is to be shared and interpreted across farms and regions.

CHAPTER 8

The Future of IOT in Tomatoes

Agriculture is our wisest pursuit, because it will in the end contribute most to real wealth, good morals, and happiness

THOMAS JEFFERSON
(Jefferson n.d.)

Among legal, high-valued crops, tomatoes are among the leaders of the pack in terms of their ability to contribute to economic wealth, but Thomas Jefferson was referring to far more than dollars when he made this statement. At the bottom line, today's tomato farmers, like every other farmer in the world, big or small, play an important role in feeding the world's population but also face the opportunity and challenge of feeding the world's increasing population tomorrow as resources become increasingly constrained. The real wealth in agriculture lies in this essential role that agriculture will always play in society despite the increasing struggle many farmers face to turn a profit.

Maintaining a prosperous farm in the modern world can be challenging and requires attending to both quality and yield, particularly for high-value crops such as tomatoes. The possibilities for using sensors to support tomatoes from seed to table for both processed products and fresh consumption are seemingly endless. Yet, the budget for doing so is often far from endless. This conflict between possibilities and reality requires a comprehensive understanding of what can be sensed and what is most important to sense. At the most basic level, what is most important to be sensed and understood can be broken down into quality and yield, although these two parameters are not always mutually exclusive.

How the quality of a tomato is evaluated depends on what its final use is. Understanding what needs to come from a processing tomato is relatively easy compared to a tomato grown for fresh consumption. Certification of processing tomatoes is based on a limited sample and allows for greater numbers of defects than is the case for fresh tomatoes. For example, during inspection and certification of California's massive processing tomato operations, not all bins are sampled and from those that are, only one sample is required from each bin selected. Representative samples of these tomatoes must pass only a single binary color test. Either the tomato meets a certain color threshold or it doesn't. In addition to this basic color test, California processing tomatoes must also meet certain population thresholds. No more than 4% of tomatoes by weight can be green, where green is defined as no visible red on the surface of the tomato. No more than 8% (by weight) of a load of tomatoes can be affected by mold and no more than 2% (by weight) can be affected by worm or insect damage. Tomatoes that are unusually soft and mushy, are broken, or have 25% of the skin separated from the tomato flesh are not rejected but instead classified as limited use. Soluble solids content and pH are also measured and recorded during inspection but cannot serve as ground for rejecting a load (California Department of Food and Agriculture 1997) (Hartz et al. 2008). Workers on board mechanized harvesters in the fields manually sort what these machines harvest to ensure that inspection requirements are met. Fast, responsive sensors have the opportunity to supplement or replace these manual sorting operations by visually monitoring tomatoes for overall color (green, pink/red, mold), defects (inhomogeneities caused by worm and insect damage), and cracking. Two-dimensional imagers or cameras are best suited for these tasks. Beyond these basic visual properties, monitoring firmness may be useful for estimating SSC and detecting mushy tomatoes within a batch. Inexpensive force or load sensors installed along the bottom of the harvester collection bins can be used to avoid limited-use classifications during inspection. And, while processing tomatoes remain in the fields, awaiting harvesting, daily monitoring of pH changes using pH sensors can allow queues into processing facilities to be optimized to minimize the need for additives after the tomatoes have been processed as well as the costs associated with such additives. In the high-volume mechanized world of processing tomatoes, batches of perfect tomatoes are not necessary, but instead those batches good enough to pass inspection can be identified by color and nonuniformities on the surface of the tomato and, to a lesser extent, soluble solids content and pH. Although not particularly important to quality, pH also contributes to profit margin in processed products by directing the quantity of additive required to maintain food safety at the end of processing. Thus, basic imagers or cameras and robust, low-maintenance pH sensors are likely some of the most attractive candidates for sensing the "good-enough" processing tomato.

Understanding what is necessary for the perfect fresh tomato is quite a bit more complicated than what is involved for the processing tomato. A distinctive difference between the two is that while the goodness of processing tomatoes is measured primarily by population or batch, the perfection of fresh tomatoes is measured on a fruit-by-fruit basis. Consumers don't buy a bag of tomatoes for eating because 80% look perfect. Instead, they expect that every tomato in the bag will look perfect or near perfect. Meeting high consumer expectations may be a nuisance in some respects, but in others, it provides many an opportunity for sensors to jump in and save the day.

What should sensors do for fresh tomatoes? The answer to this question is different depending on where in the supply chain one looks. Far in advance of harvest, image sensors, whether configured as conventional two-dimensional, hyperspectral, or multispectral cameras, can provide essential information regarding insect damage or disease that is critical to avoiding a disastrous or otherwise poor harvest. Closer to harvest, knowing how much red (or other ripe color) lies in wait in the fields (or greenhouses) is important to determining optimal time(s) to harvest. During harvest, sensing color and firmness in real time can optimize selection and minimize fruit discarded due to imperfect manual picking practices. Postharvest, sensing and monitoring color, firmness, and irregularities for individual fruit can help to guide treatment, transport, and storage practices to maximize the amount of fruit that ultimately reach the consumer. While all these data are important to optimizing harvest and handling practices, the opportunities for sensors extend much farther than this.

The IOT has a unique opportunity to solidify the links among all these sensed parameters and what matters most to the fresh tomato consumer—taste and flavor. Using a citizen science or other crowdsensing approach, it is possible for the consumer to get in on the tomato growing action. By providing images (photos) of tomatoes, the ID of their farm of origin and variety, and impressions of flavor and taste through a smart phone or similar IOT-enabled device, consumers can provide invaluable feedback to farmers. This feedback can enable the farmer to associate trends in pH as well as soluble solids and sugar content for different tomato varieties with what ultimately matters in the world of fresh tomatoes—taste and flavor.

Historically, quality has been compromised for yield, but this need not be the case. Selective breeding of tomatoes has traditionally centered on increased yields, but the corresponding loss of taste was more of an accident than an intentional genetic selection. With detailed information about taste and flavor from individual tomatoes, selective breeding and genetic alteration of tomatoes can proceed with both yield and flavor in mind, ultimately finding the best in both worlds on a region-by-region basis.

While the preceding chapters have focused on the tomato itself, there are ample opportunities for sensors to make a difference in related areas of tomato farming as well. Sensors can monitor air moisture and temperature, weather patterns, soil pH, fertilizer content, irrigation water quality, insect activity, and many other parameters that all play a role in contributing to a healthy plant and, in the end, to a healthy number of healthy tomatoes that keep farming operations in business and thriving.

> Without sensors, there is no Internet of Things.
>
> (Rogers Corporation 2016)

Although the right sensors alone do not make the IOT successful, without sensors, there is certainly greatly reduced opportunity for the IOT to thrive. This is particularly true in crop agriculture, where a myriad of characteristics associated with a particular fruit or vegetable play into whether the crop will be successful. The preceding chapters have only scratched the surface in terms of how sensors configured for and integrated into the IOT can support greater quality and yield in tomato farming. Although the possibilities for sensors in agricultural IOT are certainly not endless because of tight budgets and other resource constraints, there is definitely room for carefully chosen sensors and well-designed IOT networks to support, supplement, and enhance the wonderful world of tomatoes.

References

Albarracin, C. A., B. C. Fuqua, J. L. Evans, and I. D. Goldfine. 2008. "Chromium Picolinate and Biotin Combination Improves Glucose Metabolism in Treated, Uncontrolled Overweight to Obese Patients with Type 2 Diabetes." *Diabetes/Metabolism Research and Reviews* 24 (1): 41–51. https://doi.org/10.1002/dmrr.755.

Almassri, A. M. M., W. Z. W. Hasan, S. A. Ahmad, and A. J. Ishak. 2013. "A Sensitivity Study of Piezoresistive Pressure Sensor for Robotic Hand." In: *RSM 2013 IEEE Regional Symposium on Micro and Nanoelectronics*, 394–397. Langkawi, Malaysia: IEEE.

American Kidney Fund. 2015. 2015 Kidney Disease Statistics. http://www.kidneyfund.org/assets/pdf/kidney-disease-statistics.pdf.

Anthon, G. E., M. LeStrange, and D. M Barrett. 2011. "Changes in pH, Acids, Sugars and Other Quality Parameters during Extended Vine Holding of Ripe Processing Tomatoes." *Journal of the Science of Food and Agriculture* 91 (7): 1175–1181. https://doi.org/10.1002/jsfa.4312.

Arias, R., T.-C. Lee, L. Logendra, and H. Janes. 2000. "Correlation of Lycopene Measured by HPLC with the L*, A*, B* Color Readings of a Hydroponic Tomato and the Relationship of Maturity with Color and Lycopene Content." *Journal of Agricultural and Food Chemistry* 48 (5): 1697–1702.

Arshak, A., E. Gill, K. Arshak, O. Korostynska, and C. Cunniffe. 2007. "Drop-Coated Polyaniline Composite Conductimetric pH Sensors." In: *30th Intnl Spring Seminar on Electronics Technology*, 213–218. IEEE.

Bai, H., and G. Shi. 2007. "Gas Sensors Based on Conducting Polymers." *Sensors* 7 (3): 267–307.

Barceloux, D. G. 2009. "Potatoes, Tomatoes, and Solanine Toxicity (*Solanum Tuberosum L., Solanum Lycopersicum L.*)." *Disease-a-Month, Foodborne and Microbial Toxins, Part II: Staples and Spices* 55 (6): 391–402. https://doi.org/10.1016/j.disamonth.2009.03.009.

Barlian, A. A., W.-T. Park, J. R. Mallon, A. J. Rastegar, and B. L. Pruitt. 2009. "Review: Semiconductor Piezoresistance for Microsystems." *Proceedings of the IEEE Institute of Electrical and Electronics Engineers* 97 (3): 513–552. https://doi.org/10.1109/JPROC.2009.2013612.

Barreiro, P., A. Moya, M. Ruiz-Altisent, A. C. Agulheiro, F. J. Homer, I. García-Ramos, and G. Moreda. 2002. "On-Line Segregation of Soft Olives by Means of Differences in Rebound Trajectories." In: *Proc Intnl Conf Agricultural Engineering*. Budapest, Hungary.

Barrett, D. M., J. C. Beaulieu, and R. Shewfelt. 2010. "Color, Flavor, Texture, and Nutritional Quality of Fresh-Cut Fruits and Vegetables: Desirable Levels, Instrumental and Sensory Measurement, and the Effects of Processing." *Critical Reviews in Food Science and Nutrition* 50 (5): 369–389.

Batu, A. 2004. "Determination of Acceptable Firmness and Colour Values of Tomatoes." *Journal of Food Engineering* 61 (3): 471–475.

Berdai, M. A., S. Labib, K. Chetouani, and M. Harandou. 2012. "Atropa Belladonna Intoxication: A Case Report." *The Pan African Medical Journal* 11 (April). https://www.ncbi.nlm.nih.gov/pmc/articles/PMC3361210/.

Bertin, N., and M. Génard. 2018. "Tomato Quality as Influenced by Preharvest Factors." *Scientia Horticulturae* 233: 264–276.

Bertin, N., S. Guichard, C. Leonardi, J. J. Longuenesse, D. Langlois, and B. Navez. 2000. "Seasonal Evolution of the Quality of Fresh Glasshouse Tomatoes under Mediterranean Conditions, as Affected by Air Vapour Pressure Deficit and Plant Fruit Load." *Annals of Botany* 85 (6): 741–750.

Bezilla, M. L. n.d. "America's Tomato Crush." Accessed November 12, 2018. https://www.psu.edu/feature/2013/07/26/americas-tomato-crush.

Bhowmik, D., K. P. S. Kumar, S. Paswan, and S. Srivastava. 2012. "Tomato—A Natural Medicine and Its Health Benefits." *Journal of Pharmacognosy and Phytochemistry* 1 (1): 33–43.

Block, G. 1991. "Vitamin C and Cancer Prevention: The Epidemiologic Evidence." *The American Journal of Clinical Nutrition* 53 (1): 270S–282S.

BLS, US Bureau of Labor Statistics. 2017. "Employment by Major Industry Sector." Accessed October 24, 2017. https://www.bls.gov/emp/tables/employment-by-major-industry-sector.htm.

Blum, A., M. Monir, I. Wirsansky, and S. Ben-Arzi. 2005. "The Beneficial Effects of Tomatoes." *European Journal of Internal Medicine* 16 (6): 402–404. https://doi.org/10.1016/j.ejim.2005.02.017.

Boisen, A., and T. Thundat. 2009. "Design & Fabrication of Cantilever Array Biosensors." *Materials Today* 12 (9): 32–38. https://doi.org/10.1016/S1369-7021(09)70249-4.

Boom California. 2013. "The Tomato Harvester." *Boom California (Blog)*. Accessed June 24, 2013. https://boomcalifornia.com/2013/06/24/thinking-through-the-tomato-harvester/.

Butz, P., C. Hofmann, and B. Tauscher. 2005. "Recent Developments in Noninvasive Techniques for Fresh Fruit and Vegetable Internal Quality Analysis." *Journal of Food Science* 70 (9): R131–R141.

California Department of Food and Agriculture. 1997. "California Processing Tomato Inspection Program." California Processing Tomato Inspection Program. Accessed November 26, 2018. http://www.ptab.org/order.htm.

Canada, Agriculture and Agri-Food. 2017. "Statistical Overview of the Canadian Greenhouse Vegetable Industry—2016." Fact sheet. Accessed February 22, 2019. http://www.agr.gc.ca/eng/industry-markets-and-trade/canadian-agri-food-sector-intelligence/horticulture/horticulture-sector-reports/statistical-overview-of-the-canadian-greenhouse-vegetable-industry-2016/?id=1511888129369.

Cashman, K. D. 2007. "Diet, Nutrition, and Bone Health." *The Journal of Nutrition* 137 (11): 2507S–2512S. https://doi.org/10.1093/jn/137.11.2507S.

Causse, M., M. Buret, K. Robini, and P. Verschave. 2003. "Inheritance of Nutritional and Sensory Quality Traits in Fresh Market Tomato and Relation to Consumer Preferences." *Journal of Food Science* 68 (7): 2342–2350.

Causse, M., C. Friguet, C. Coiret, M. Lépicier, B. Navez, M. Lee, N. Holthuysen, F. Sinesio, E. Moneta, and S. Grandillo. 2010. "Consumer Preferences for Fresh Tomato at the European Scale: A Common Segmentation on Taste and Firmness." *Journal of Food Science* 75 (9): S531–S541.

Causse, M., V. Saliba-Colombani, L. Lecomte, P. Duffe, P. Rousselle, and M. Buret. 2002. "QTL Analysis of Fruit Quality in Fresh Market Tomato: A Few Chromosome Regions Control the Variation of Sensory and Instrumental Traits." *Journal of Experimental Botany* 53 (377): 2089–2098.

Cerovic, Z. G., J.-P. Goutouly, G. Hilbert, A. Destrac-Irvine, V. Martinon, and N. Moise. 2009. "Mapping Winegrape Quality Attributes Using Portable Fluorescence-Based Sensors." *Frutic* 9: 301–310.

Chandra, R. V., M. L. Prabhuji, D. Adinarayana Roopa, S. Ravirajan, and H. C. Kishore. 2007. "Efficacy of Lycopene in the Treatment of Gingivitis: A Randomised, Placebo-Controlled Clinical Trial." *Oral Health & Preventive Dentistry* 5 (4): 327–336.

Charles, D. 2016. "How Canada Became A Greenhouse Superpower." NPR. Accessed June 16, 2016. https://www.npr.org/sections/thesalt/2016/06/16/473526920/how-canada-became-a-greenhouse-superpower.

Chen, P., Margarita Ruiz-Altisent, and P. Barreiro. 1996. "Effect of Impacting Mass on Firmness Sensing of Fruits." *Transactions of the ASAE* 39 (3): 1019–1023.

Cherng, A.-P. 2008. "Development of a Frequency-Based Firmness Index for Solid Fruits and Vegetables of Ellipsoidal Shape." *Engineering in Agriculture, Environment and Food* 2 (1): 7–13. https://doi.org/10.11165/eaef.2.7.

Chudy, M., W. Wroblewski, and Z. Brzózka. 1999. "Towards REFET." *Sensors and Actuators B: Chemical* 57 (1–3): 47–50.

Ciampa, A., M. Teresa Dell'Abate, O. Masetti, M. Valentini, and P. Sequi. 2010. "Seasonal Chemical–Physical Changes of PGI Pachino Cherry Tomatoes Detected by Magnetic Resonance Imaging (MRI)." *Food Chemistry* 122 (4): 1253–1260. https://doi.org/10.1016/j.foodchem.2010.03.078.

Clemson University Extension. 2018. "pH Value of Common Foods and Ingredients." https://www.clemson.edu/extension/food/food2market/documents/ph_of_common_foods.pdf.

Clinton, S. K. 1998. "Lycopene: Chemistry, Biology, and Implications for Human Health and Disease." *Nutrition Reviews* 56 (2): 35–51.

Cockayne, S., J. Adamson, S. Lanham-New, M. J. Shearer, S. Gilbody, and D. J. Torgerson. 2006. "Vitamin K and the Prevention of Fractures: Systematic Review and Meta-Analysis of Randomized Controlled Trials." *Archives of Internal Medicine* 166 (12): 1256–1261.

Cui, Y., J. M. Shikany, S. Liu, Y. Shagufta, and T. E. Rohan. 2008. "Selected Antioxidants and Risk of Hormone Receptor–Defined Invasive Breast Cancers among Postmenopausal Women in the Women's Health Initiative Observational Study." *The American Journal of Clinical Nutrition* 87 (4): 1009–1018.

Davies, J. N., G. E. Hobson, and W. B. McGlasson. 1981. "The Constituents of Tomato Fruit—The Influence of Environment, Nutrition, and Genotype." *C R C Critical Reviews in Food Science and Nutrition* 15 (3): 205–280. https://doi.org/10.1080/10408398109527317.

Davis, A. R., W. W. Fish, and P. Perkins-Veazie. 2003. "A Rapid Spectrophotometric Method for Analyzing Lycopene Content in Tomato and Tomato Products." *Postharvest Biology*

and Technology 28 (3): 425–430. https://doi.org/10.1016/
S0925-5214(02)00203-X.

De Ketelaere, B., M. S. Howarth, L. Crezee, J. Lammertyn, K. Viaene,
I. Bulens, and J. De Baerdemaeker. 2006. "Postharvest Firmness
Changes as Measured by Acoustic and Low-Mass Impact Devices:
A Comparison of Techniques." *Postharvest Biology and Technology*
41 (3): 275–284. https://doi.org/10.1016/j.postharvbio.2006.04.008.

Desjardins, J. 2014. "The World's Most Valuable Cash Crop."
Visual Capitalist. Accessed November 10, 2014. http://www.
visualcapitalist.com/the-worlds-most-valuable-cash-crop/.

Doll, J. C., and B. L. Pruitt. 2013. *Piezoresistor Design and Applications.*
Springer.

Dorais, M., D. L. Ehret, and A. P. Papadopoulos. 2008. "Tomato
(*Solanum Lycopersicum*) Health Components: From the Seed to
the Consumer." *Phytochemistry Reviews* 7 (2): 231. https://doi.
org/10.1007/s11101-007-9085-x.

Duma, M., I. Alsina, L. Dubova, and I. Erdberga. 2015. *Chemical
Composition of Tomatoes Depending on the Stage of Ripening.*
Vol. 66. https://doi.org/10.5755/j01.ct.66.1.12053.

Eschner, K. 2017. "Tomatoes Have Legally Been Vegetables Since
1893." Smithsonian. Accessed May 10, 2017. https://www.
smithsonianmag.com/smart-news/even-supreme-court-maintains-
tomato-vegetable-180963133/.

FAO, Food and Agriculture Organization of the United Nations. 2017a.
"SOFA 2017—The State of Food and Agriculture." 2017. http://
www.fao.org/state-of-food-agriculture/en/.

FAO, Food and Agriculture Organization of the United Nations. 2017b.
The Future of Food and Agriculture: Trends and Challenges. Rome:
Food and Agriculture Organization of the United Nations.

FAOSTAT, Food and Agriculture Organization of the United Nations.
2018. "Crops Data." Accessed November 7, 2018. http://www.fao.
org/faostat/en/#data.

Fernandez-Mejia, C. 2005. "Pharmacological Effects of Biotin." *The
Journal of Nutritional Biochemistry* 16 (7): 424–427. https://doi.
org/10.1016/j.jnutbio.2005.03.018.

Food and Agriculture Organization of the United Nations (FAO). 2009.
"Global Agriculture Towards 2050." Accessed September 18, 2018.
http://www.fao.org/fileadmin/templates/wsfs/docs/Issues_papers/
HLEF2050_Global_Agriculture.pdf.

Frary, A., D. Göl, D. Keleş, B. Ökmen, H. Pınar, H. Ö. Şığva, A.
Yemenicioğlu, and S. Doğanlar. 2010. "Salt Tolerance in *Solanum*

Pennellii: Antioxidant Response and Related QTL." *BMC Plant Biology* 10 (1): 58. https://doi.org/10.1186/1471-2229-10-58.

FREP, Fertilizer Research and Education Program. 2013. "Production of Processing Tomatoes in California." https://apps1.cdfa.ca.gov/FertilizerResearch/docs/Tomato_Production_CA.pdf.

Gallicchio, L., K. Boyd, G. Matanoski, X. (Grant) Tao, L. Chen, T. K. Lam, M. Shiels et al. 2008. "Carotenoids and the Risk of Developing Lung Cancer: A Systematic Review." *The American Journal of Clinical Nutrition* 88 (2): 372–383. https://doi.org/10.1093/ajcn/88.2.372.

García-Ramos, F. J., C. Valero, I. Homer, J. Ortiz-Cañavate, and M. Ruiz-Altisent. 2005. "Non-Destructive Fruit Firmness Sensors: A Review." *Spanish Journal of Agricultural Research* 3 (1): 61–73.

Gautier, H., V. Diakou-Verdin, C. Bénard, M. Reich, M. Buret, F. Bourgaud, J. L. Poëssel, C. Caris-Veyrat, and M. Génard. 2008. "How Does Tomato Quality (Sugar, Acid, and Nutritional Quality) Vary with Ripening Stage, Temperature, and Irradiance?" *Journal of Agricultural and Food Chemistry* 56 (4): 1241–1250. https://doi.org/10.1021/jf072196t.

Geleijnse, J. M., C. Vermeer, D. E. Grobbee, L. J. Schurgers, M. H. J. Knapen, I. M. Van Der Meer, A. Hofman, and J. C. M. Witteman. 2004. "Dietary Intake of Menaquinone Is Associated with a Reduced Risk of Coronary Heart Disease: The Rotterdam Study." *The Journal of Nutrition* 134 (11): 3100–3105.

Gentilcore, D. 2010. *Pomodoro!: A History of the Tomato in Italy.* Columbia University Press.

Geohas, J., A. Daly, V. Juturu, M. Finch, and J. R. Komorowski. 2007. "Chromium Picolinate and Biotin Combination Reduces Atherogenic Index of Plasma in Patients with Type 2 Diabetes Mellitus: A Placebo-Controlled, Double-Blinded, Randomized Clinical Trial." *The American Journal of the Medical Sciences* 333 (3): 145–153. https://doi.org/10.1097/MAJ.0b013e318031b3c9.

Georgé, S., F. Tourniaire, H. Gautier, P. Goupy, E. Rock, and C. Caris-Veyrat. 2011. "Changes in the Contents of Carotenoids, Phenolic Compounds and Vitamin C during Technical Processing and Lyophilisation of Red and Yellow Tomatoes." *Food Chemistry* 124 (4): 1603–1611. https://doi.org/10.1016/j.foodchem.2010.08.024.

Gill, E., A. Arshak, K. Arshak, and O. Korostynska. 2008a. "Conductimetric PH Sensor Based on Novel Conducting Polymer Composite Thick Films." In: *2008 31st International Spring Seminar on Electronics Technology*, 478–483. Budapest, Hungary: IEEE. https://doi.org/10.1109/ISSE.2008.5276613.

Gill, E., K. Arshak, A. Arshak, and O. Korostynska. 2008b. "Mixed Metal Oxide Films as pH Sensing Materials." *Microsystem Technologies* 14 (4): 499–507. https://doi.org/10.1007/s00542-007-0435-9.

Giovannucci, E. n.d. "Tomatoes, Tomato-Based Products, Lycopene, and Cancer: Review of the Epidemiologic Literature." *JNCI: Journal of the National Cancer Institute* 91 (4): 317–331.

Giovannucci, E., E. B. Rimm, Y. Liu, M. J. Stampfer, and W. C. Willett. 2002. "A Prospective Study of Tomato Products, Lycopene, and Prostate Cancer Risk." *JNCI: Journal of the National Cancer Institute* 94 (5): 391–398. https://doi.org/10.1093/jnci/94.5.391.

Goel, N., and P. Sehgal. 2015. "Fuzzy Classification of Pre-Harvest Tomatoes for Ripeness Estimation–An Approach Based on Automatic Rule Learning Using Decision Tree." *Applied Soft Computing* 36: 45–56.

Grand Voyage Italy. 2016. "How the Tomato Became Part of Italian Culture." Grand Voyage Italy. Accessed May 3, 2016. http://www.grandvoyageitaly.com/1/post/2016/05/how-the-tomato-became-part-of-italian-culture.html.

Guan, Z., F. Wu, and T. Biswas. 2017. "The US Tomato Industry: An Overview of Production and Trade." September 2017. http://edis.ifas.ufl.edu/fe1027.

Guichard, S., C. Gary, C. Leonardi, and N. Bertin. 2005. "Analysis of Growth and Water Relations of Tomato Fruits in Relation to Air Vapor Pressure Deficit and Plant Fruit Load." *Journal of Plant Growth Regulation* 24 (3): 201.

Gundberg, C. M., J. B. Lian, and S. L. Booth. 2012. "Vitamin K-Dependent Carboxylation of Osteocalcin: Friend or Foe?" *Advances in Nutrition* 3 (2): 149–157.

Gutheil, R. A., L. G. Price, and B. G. Swanson. 1980. "pH, Acidity, and Vitamin C Content of Fresh and Canned Homegrown Washington Tomatoes." *Journal of Food Protection* 43 (5): 366–369.

Hahn, F. 2002. "AE—Automation and Emerging Technologies: Multi-Spectral Prediction of Unripe Tomatoes." *Biosystems Engineering* 81 (2): 147–155. https://doi.org/10.1006/bioe.2001.0035.

Handwerk, B. 2017. "The Quest to Return Tomatoes to Their Full-Flavored Glory." Smithsonian. Accessed January 26, 2017. https://www.smithsonianmag.com/science-nature/geneticists-quest-return-tomatoes-full-flavored-glory-180961933/.

Hanna, Instruments. 2015. "Determining Tomato Maturity by Measuring pH." Hanna Instruments. Accessed June 17, 2015. http://blog.hannainst.com/determining-tomato-maturity-by-measuring-ph/.

Harker, F. R., J. Maindonald, S. H. Murray, F. A. Gunson, I. C. Hallett, and S. B. Walker. 2002. "Sensory Interpretation of Instrumental Measurements 1: Texture of Apple Fruit." *Postharvest Biology and Technology* 24 (3): 225–239.

Hartz, T., G. Miyao, J. Mickler, M. Lestrange, S. Stoddard, J. Nuñez, and B. Aegerter. 2008. *Processing Tomato Production in California.* University of California, Agriculture and Natural Resources. https://doi.org/10.3733/ucanr.7228.

health.gov. 2015. "Report Index—2015 Advisory Report—Health.Gov." https://health.gov/dietaryguidelines/2015-scientific-report/.

Heflebower, R., and C. Washburn. 2010. "The Influence of Different Tomato Varieties on Acidity as It Relates to Home Canning." *Journal of Extension* 48 (6). https://www.joe.org/joe/2010december/rb6.php.

Hobson, G. E., P. Adams, and T. J. Dixon. 1983. "Assessing the Colour of Tomato Fruit during Ripening." *Journal of the Science of Food and Agriculture* 34 (3): 286–292.

Hoppner, K., B. Lampi, and E. O'Grady. 1994. "Biotin Content in Vegetables and Nuts Available on the Canadian Market." *Food Research International* 27 (5): 495–497. https://doi.org/10.1016/0963-9969(94)90245-3.

Hristoforou, E., and D. S. Vlachos. 2014. "The Effect of Sheet Resistivity and Storage Conditions on Sensitivity of RuO2 Based pH Sensors." April 2014. https://www.scientific.net/KEM.605.457.

Huang, C.-T., C.-L. Shen, C.-F. Tang, and S.-H. Chang. 2008. "A Wearable Yarn-Based Piezo-Resistive Sensor." *Sensors and Actuators A: Physical* 141 (2): 396–403. https://doi.org/10.1016/j.sna.2007.10.069.

Insights, MIT Technology Review. 2016. "IoT: The Internet of Tomatoes." MIT Technology Review. Accessed September 18, 2018. https://www.technologyreview.com/s/601793/iot-the-internet-of-tomatoes/.

Jackman, R. L., A. G. Marangoni, and D. W. Stanley. 1990. "Measurement of Tomato Fruit Firmness." *HortScience* 25 (7): 781–783.

Janata, J. 2003. "Electrochemical Microsensors." *Proceedings of the IEEE* 91 (6): 864–869. https://doi.org/10.1109/JPROC.2003.813576.

Jefferson. n.d. "Founders Online: From Thomas Jefferson to George Washington, 14 August 1787." Accessed November 26, 2018. http://founders.archives.gov/documents/Jefferson/01-12-02-0040.

Jeong, N.-H., E.-S. Song, J.-M. Lee, K.-B. Lee, M.-K. Kim, J.-E. Cheon, J.-K. Lee et al. 2009. "Plasma Carotenoids, Retinol and Tocopherol Levels and the Risk of Ovarian Cancer." *Acta Obstetricia et Gynecologica Scandinavica* 88 (4): 457–462. https://doi.org/10.1080/00016340902807215.

Kennedy, D. O. 2016. "B Vitamins and the Brain: Mechanisms, Dose and Efficacy—A Review." *Nutrients* 8 (2). https://doi.org/10.3390/nu8020068.

Khan, M. I., K. Mukherjee, R. Shoukat, and H. Dong. 2017. "A Review on pH Sensitive Materials for Sensors and Detection Methods." *Microsystem Technologies* 23 (10): 4391–4404. https://doi.org/10.1007/s00542-017-3495-5.

Khankaew, S., W. Boonsupthip, C. Pechyen, and P. Suppakul. n.d. "Screening of Naturally-Derived pH Dyes from Plant Extract Powders as Colorimetric Bio-Indicator and Possible Application in Intelligent Packaging." In: *26th IAPRI Symposium on Packaging*. Esspoo, Finland.

Kim, T.-J., J.-S. Byun, H. S. Kwon, and D.-Y. Kim. 2018. "Cellular Toxicity Driven by High-Dose Vitamin C on Normal and Cancer Stem Cells." *Biochemical and Biophysical Research Communications*, February. https://doi.org/10.1016/j.bbrc.2018.02.083.

Kristal, A. R., K. B. Arnold, J. M. Schenk, M. L. Neuhouser, P. Goodman, D. F. Penson, and I. M. Thompson. 2008. "Dietary Patterns, Supplement Use, and the Risk of Symptomatic Benign Prostatic Hyperplasia: Results from the Prostate Cancer Prevention Trial." *American Journal of Epidemiology* 167 (8): 925–934. https://doi.org/10.1093/aje/kwm389.

Larsson, S. C., L. Bergkvist, I. Näslund, J. Rutegaard, and A. Wolk. 2007. "Vitamin A, Retinol, and Carotenoids and the Risk of Gastric Cancer: A Prospective Cohort Study." *The American Journal of Clinical Nutrition* 85 (2): 497–503.

Lee, R. L., and J. Hernández-Andrés. 2005. "Colors of the Daytime Overcast Sky." *Applied Optics* 44 (27): 5712. https://doi.org/10.1364/AO.44.005712.

LeHoullier, C.g. 2014. *Epic Tomatoes: How to Select and Grow the Best Varieties of All Time*. Storey Publishing, LLC.

Leung, E. Y. L., J. E. M. Crozier, D. Talwar, D. St. J. O'Reilly, R. F. McKee, P. G. Horgan, and D. C. McMillan. 2008. "Vitamin Antioxidants, Lipid Peroxidation, Tumour Stage, the Systemic Inflammatory Response and Survival in Patients with Colorectal Cancer." *International Journal of Cancer* 123 (10): 2460–2464.

Li, B., J. Lecourt, G. Bishop, B. Li, J. Lecourt, and G. Bishop. 2018. "Advances in Non-Destructive Early Assessment of Fruit Ripeness towards Defining Optimal Time of Harvest and Yield Prediction—A Review." *Plants* 7 (1): 3. https://doi.org/10.3390/plants7010003.

Li, D., A. Radulescu, R. T. Shrestha, M. Root, A. B. Karger, A. A. Killeen, J. S. Hodges, S.-L. Fan, A. Ferguson, and U. Garg. 2017. "Association of

Biotin Ingestion with Performance of Hormone and Nonhormone Assays in Healthy Adults." *Jama* 318 (12): 1150–1160.

Li, Y., and H. E. Schellhorn. 2007. "New Developments and Novel Therapeutic Perspectives for Vitamin C." *The Journal of Nutrition* 137 (10): 2171–2184.

Liu, C., W. Liu, W. Chen, J. Yang, and L. Zheng. 2015. "Feasibility in Multispectral Imaging for Predicting the Content of Bioactive Compounds in Intact Tomato Fruit." *Food Chemistry* 173 (April): 482–488. https://doi.org/10.1016/j.foodchem.2014.10.052.

López Camelo, A. F., and P. A. Gómez. 2004. "Comparison of Color Indexes for Tomato Ripening." *Horticultura Brasileira* 22 (3): 534–537. https://doi.org/10.1590/S0102-05362004000300006.

Lorente, D., N. Aleixos, J. Gómez-Sanchis, S. Cubero, O. L. García-Navarrete, and J. Blasco. 2012. "Recent Advances and Applications of Hyperspectral Imaging for Fruit and Vegetable Quality Assessment." *Food and Bioprocess Technology* 5 (4): 1121–1142. https://doi.org/10.1007/s11947-011-0725-1.

Manyika, J., S. Ramaswamy, S. Khanna, H. Sarrazin, G. Pinkus, G. Sethupathy, and A. Yaffe. 2015. "Digital America: A Tale of the Haves and Have-Mores." December 2015. https://www.mckinsey.com/industries/high-tech/our-insights/digital-america-a-tale-of-the-haves-and-have-mores.

Marsh, A. 2018. "John Deere and the Birth of Precision Agriculture." IEEE Spectrum: Technology, Engineering, and Science News. Accessed February 28, 2018. https://spectrum.ieee.org/tech-history/silicon-revolution/john-deere-and-the-birth-of-precision-agriculture.

Massot, C., M. Génard, R. Stevens, and H. Gautier. 2010. "Fluctuations in Sugar Content Are Not Determinant in Explaining Variations in Vitamin C in Tomato Fruit." *Plant Physiology and Biochemistry* 48 (9): 751–757.

Monti, L. M. 1979. "The Breeding of Tomatoes for Peeling." In: *Symposium on Production of Tomatoes for Processing* 100, 341–354.

Naidu, K. A. 2003. "Vitamin C in Human Health and Disease Is Still a Mystery? An Overview." *Nutrition Journal* 2 (1): 7. https://doi.org/10.1186/1475-2891-2-7.

National Institutes of Health (NIH). 2018e. "Office of Dietary Supplements—Vitamin C." Accessed September 17, 2018. https://ods.od.nih.gov/factsheets/VitaminC-HealthProfessional/.

National Institutes of Health (NIH). 2018f. "Office of Dietary Supplements—Vitamin K." Accessed September 17, 2018. https://ods.od.nih.gov/factsheets/VitaminK-HealthProfessional/.

National Institutes of Health (NIH). 2018a. "Potassium Fact Sheet for the Health Professional." Accessed August 20, 2018. https://ods.od.nih.gov/factsheets/Potassium-HealthProfessional/.

National Institutes of Health (NIH). 2018b. "Vitamin B7 Fact Sheet for Health Professionals." Accessed September 17, 2018. https://ods.od.nih.gov/factsheets/Biotin-HealthProfessional/.

National Institutes of Health (NIH). 2018c. "Vitamin C Fact Sheet for Health Professionals." Accessed September 18, 2018. https://ods.od.nih.gov/factsheets/VitaminC-HealthProfessional/.

National Institutes of Health (NIH). 2018d. "Vitamin K Fact Sheet for Health Professionals." Accessed September 26, 2018. https://ods.od.nih.gov/factsheets/VitaminK-HealthProfessional/.

Niño-Medina, G., J. C. Rivera-Castro, J. A. Vidales-Contreras, H. Rodriguez-Fuentes, and A. I. Luna-Maldonado. 2013. "PhysicoChemical Parameters for Obtaining Prediction Models in the Postharvest Quality of Tomatoes (*Solanum Lycopersicum L.*)." *Trans. MLDM* 6 (2): 81–91.

Nouraie, M., P. Pietinen, F. Kamangar, S. M. Dawsey, C. C. Abnet, D. Albanes, J. Virtamo, and P. R. Taylor. 2005. "Fruits, Vegetables, and Antioxidants and Risk of Gastric Cancer among Male Smokers." *Cancer Epidemiology and Prevention Biomarkers* 14 (9): 2087–2092. https://doi.org/10.1158/1055-9965.EPI-05-0038.

NPR Staff. 2011. "The Troubled History Of The Supermarket Tomato." NPR. Accessed July 9, 2011. https://www.npr.org/2011/07/09/137623954/the-troubled-history-of-the-supermarket-tomato.

OEC (Observatory of Economic Complexity). n.d. "Tomatoes." Accessed November 7, 2018. https://atlas.media.mit.edu/en/profile/hs92/0702/.

Pacheco-Alvarez, D., R. S. Solórzano-Vargas, and A. Del Río. 2002. "Biotin in Metabolism and Its Relationship to Human Disease." *Archives of Medical Research* 33 (5): 439–447. https://doi.org/10.1016/S0188-4409(02)00399-5.

Park, D. M., S. A. White, and N. Menchyk. 2014. "Assessing Irrigation Water Quality for pH, Salts, & Alkalinity." *Journal of Extension* 52 (6). https://www.joe.org/joe/2014december/tt8.php.

Parr, B., J. K. Bond, and T. Minor. 2018. "Vegetables and Pulses Outlook." USDA (United States Department of Agriculture). https://www.ers.usda.gov/webdocs/publications/88712/vgs-360.pdf?v=0.

Penas, E. J., and D. T. Lindgren. 1990. "G90-945 A Gardener's Guide for Soil and Nutrient Management in Growing Vegetables." Universtiy of Nebraska Lincoln. https://digitalcommons.unl.edu/cgi/viewcontent.cgi?article=2011&context=extensionhist.

Pennsylvania State University. n.d. "Acid Base Indicators." Accessed October 28, 2018. http://chemistry.bd.psu.edu/jircitano/abindic.html.

Peters, U., M. F. Leitzmann, N. Chatterjee, Y. Wang, D. Albanes, E. P. Gelmann, M. D. Friesen, E. Riboli, and R. B. Hayes. 2007. "Serum Lycopene, Other Carotenoids, and Prostate Cancer Risk: A Nested Case-Control Study in the Prostate, Lung, Colorectal, and Ovarian Cancer Screening Trial." *Cancer Epidemiology and Prevention Biomarkers* 16 (5): 962–968. https://doi.org/10.1158/1055-9965. EPI-06-0861.

Pitts, M., R. Cavalieri, S. Drake, and J. Fellman. 1997. "Evaluating Apple Firmness Sensors." Washington State University Tree Fruit Research & Extension Center. December 1997. http://postharvest. tfrec.wsu.edu/pages/J8I4C.

Pu, Y.-Y., Y.-Z. Feng, and D.-W. Sun. 2015. "Recent Progress of Hyperspectral Imaging on Quality and Safety Inspection of Fruits and Vegetables: A Review." *Comprehensive Reviews in Food Science and Food Safety* 14 (2): 176–188.

Qin, Y., H.-J. Kwon, M. M. R. Howlader, and M. Jamal Deen. 2015. "Microfabricated Electrochemical pH and Free Chlorine Sensors for Water Quality Monitoring: Recent Advances and Research Challenges." *RSC Advances* 5 (85): 69086–69109. https://doi. org/10.1039/C5RA11291E.

Reddi, A., B. DeAngelis, O. Frank, N. Lasker, and H. Baker. 1988. "Biotin Supplementation Improves Glucose and Insulin Tolerances in Genetically Diabetic KK Mice." *Life Sciences* 42 (13): 1323–1330. https://doi.org/10.1016/0024-3205(88)90226-3.

Richter, A., G. Paschew, S. Klatt, J. Lienig, K.-F. Arndt, and H.-J. P. Adler. 2008. "Review on Hydrogel-Based pH Sensors and Microsensors." *Sensors* 8 (1): 561–581.

Ripoll, J., L. Urban, B. Brunel, and N. Bertin. 2016. "Water Deficit Effects on Tomato Quality Depend on Fruit Developmental Stage and Genotype." *Journal of Plant Physiology* 190 (January): 26–35. https://doi.org/10.1016/j.jplph.2015.10.006.

Ripoll, J., L. Urban, M. Staudt, F. Lopez-Lauri, L. P. R. Bidel, and N. Bertin. 2014. "Water Shortage and Quality of Fleshy Fruits— Making the Most of the Unavoidable." *Journal of Experimental Botany* 65 (15): 4097–4117.

Rogers Corporation. 2016. "Without Sensors, There Is No Internet of Things (IoT)—Rogers Corporation Blog." Accessed November 26, 2018. http://blog.rogerscorp.com/2016/07/31/ without-sensors-there-is-no-internet-of-things-iot/.

Royer, C. A. 1995. "Fluorescence Spectroscopy." *Methods in Molecular Biology (Clifton, N.J.)* 40: 65–89. https://doi.org/10.1385/0-89603-301-5:65.

Rutgers, New Jersey Agricultural Experiment Station. n.d. "Tomato Varieties." Accessed November 26, 2018. https://njaes.rutgers.edu/tomato-varieties/.

Saad, A. M., A. Ibrahim, and N. El-Bialee. 2016. "Internal Quality Assessment of Tomato Fruits Using Image Color Analysis." *Agricultural Engineering International: CIGR Journal* 18 (1): 339–352.

Sahni, S., M. T. Hannan, J. Blumberg, L. Adrienne Cupples, D. P. Kiel, and K. L. Tucker. 2009. "Protective Effect of Total Carotenoid and Lycopene Intake on the Risk of Hip Fracture: A 17-Year Follow-up from the Framingham Osteoporosis Study." *Journal of Bone and Mineral Research* 24 (6): 1086–1094.

Salim, A., and S. Lim. 2017. "Review of Recent Inkjet-Printed Capacitive Tactile Sensors." *Sensors* 17 (11): 2593.

Satia, J. A., A. Littman, C. G. Slatore, J. A. Galanko, and E. White. 2009. "Long-Term Use of β-Carotene, Retinol, Lycopene, and Lutein Supplements and Lung Cancer Risk: Results from the Vitamins and Lifestyle (VITAL) Study." *American Journal of Epidemiology* 169 (7): 815–828.

Schaller, R. R. 1997. "Moore's Law: Past, Present and Future." *IEEE Spectrum* 34 (6): 52–59. https://doi.org/10.1109/6.591665.

Schouten, R. E., T. P. M. Huijben, L. M. M. Tijskens, and O. van Kooten. 2007. "Modelling Quality Attributes of Truss Tomatoes: Linking Colour and Firmness Maturity." *Postharvest Biology and Technology* 45 (3): 298–306.

Schurgers, L. J. 2013. "Vitamin K: Key Vitamin in Controlling Vascular Calcification in Chronic Kidney Disease." *Kidney International* 83 (5): 782–784.

Sea-Bird Scientific. 2018. "Comparing ISFET and Glass Electrode PH Sensors." 2018. https://www.seabird.com/cms-portals/seabird_com/cms/documents/casestudy-isfet-glass_0.pdf.

Seymour, G. B., K. Manning, E. M. Eriksson, A. H. Popovich, and G. J. King. 2002. "Genetic Identification and Genomic Organization of Factors Affecting Fruit Texture." *Journal of Experimental Botany* 53 (377): 2065–2071.

Shea, M. Kyla, C. J. O'Donnell, U. Hoffmann, G. E. Dallal, B. Dawson-Hughes, J. M. Ordovas, P. A. Price, M. K. Williamson, and S. L. Booth. 2009. "Vitamin K Supplementation and Progression of Coronary Artery Calcium in Older Men and Women–." *The American Journal of Clinical Nutrition* 89 (6): 1799–1807.

Shen, Y.-C., S.-L. Chen, and C.-K. Wang. 2007. "Contribution of Tomato Phenolics to Antioxidation and Down-Regulation of Blood Lipids." *Journal of Agricultural and Food Chemistry* 55 (16): 6475–6481.

Shewfelt, R. L. 2000. "Fruit and Vegetable Quality." In: *Fruit and Vegetable Quality: An Integrated View*, edited by Robert L. Shewfelt and Bernhard Bruckner. CRC Press.

Shmulevich I., N. Galili, and D. Rosenfeld. 1996. "Detection of Fruit Firmness by Frequency Analysis." *Transactions of the ASAE* 39 (3): 1047–1055. https://doi.org/10.13031/2013.27595.

Simic, M., L. Manjakkal, K. Zaraska, G. M. Stojanovic, and R. Dahiya. 2017. "TiO_2-Based Thick Film pH Sensor." *IEEE Sensors Journal* 17 (2): 248–255. https://doi.org/10.1109/JSEN.2016.2628765.

Smith, A. F. 2001. *The Tomato in America: Early History, Culture, and Cookery*. University of Illinois Press.

Smith, K. A. 2013. "Why the Tomato Was Feared in Europe for More Than 200 Years." Accessed June 18, 2013. https://www.smithsonianmag.com/arts-culture/why-the-tomato-was-feared-in-europe-for-more-than-200-years-863735/.

Spijkman, M., E. C. P. Smits, J. F. M. Cillessen, F. Biscarini, P. W. M. Blom, and D. M. de Leeuw. 2011. "Beyond the Nernst-Limit with Dual-Gate ZnO Ion-Sensitive Field-Effect Transistors." *Applied Physics Letters* 98 (4): 043502. https://doi.org/10.1063/1.3546169.

Stahl, W., U. Heinrich, S. Wiseman, O. Eichler, H. Sies, and H. Tronnier. 2001. "Dietary Tomato Paste Protects against Ultraviolet Light–Induced Erythema in Humans." *The Journal of Nutrition* 131 (5): 1449–1451.

Statista. n.d. "U.S. Fresh Tomato Consumption per Capita, 2017 | Statistic." Statista. Accessed November 12, 2018. https://www.statista.com/statistics/257302/per-capita-consumption-of-fresh-tomatoes-in-the-us/.

Steinmetz, V., M. Crochon, V. Bellon Maurel, J. L. García Fernández, P. B. Elorza, and L. Verstreken. 1996. "Sensors for Fruit Firmness Assessment: Comparision and Fusion." *Journal of Agricultural Engineering Research* 64 (May): 15–27.

Story, E. N., R. E. Kopec, S. J. Schwartz, and G. Keith Harris. 2010. "An Update on the Health Effects of Tomato Lycopene." *Annual Review of Food Science and Technology* 1. https://doi.org/10.1146/annurev.food.102308.124120.

Takahashi, N., H. Maki, H. Nishina, and K. Takayama. 2013. "Evaluation of Tomato Fruit Color Change with Different Maturity Stages and Storage Temperatures Using Image Analysis." *IFAC Proceedings Volumes* 46 (4): 147–149.

Tandon, K. S., E. A. Baldwin, J. W. Scott, and R. L. Shewfelt. 2003. "Linking Sensory Descriptors to Volatile and Nonvolatile Components of Fresh Tomato Flavor." *Journal of Food Science* 68 (7): 2366–2371. https://doi.org/10.1111/j.1365-2621.2003.tb05774.x.

Teledyne Dalsa. n.d. "CCD vs CMOS: Which Is Better? It's Complicated." Accessed September 9, 2018. https://www.teledynedalsa.com/en/learn/knowledge-center/ccd-vs-cmos/.

Thusu, R.r. 2012. "The Growing World of the Image Sensors Market." *Sensors Magazine,* Accessed February 1, 2012. https://www.sensorsmag.com/embedded/growing-world-image-sensors-market.

Tran, D. T., M. L. A. T. M. Hertog, T. L. H. Tran, N. T. Quyen, B. Van de Poel, C. I. Mata, and B. M. Nicolaï. 2017. "Population Modeling Approach to Optimize Crop Harvest Strategy. The Case of Field Tomato." *Frontiers in Plant Science* 8 (April). https://doi.org/10.3389/fpls.2017.00608.

Trüeb, R. M. 2016. "Serum Biotin Levels in Women Complaining of Hair Loss." *International Journal of Trichology* 8 (2): 73–77. https://doi.org/10.4103/0974-7753.188040.

UGA Extension, University of Georgia. 2017. "Commercial Tomato Production Handbook." Accessed January 30, 2017. http://extension.uga.edu/publications/detail.html?number=B1312#Harvest.

UMMS, University of Maryland Medical System. n.d. "Kidneys and Kidney Disease." https://www.umms.org/ummc/health-services/kidney/disease.

UNODC. 2014. "UNODC (United Nations Office on World Drug Report on Drugs and Crime) World Drug Report." 2014. http://www.unodc.org/wdr2014/.

USDA, United States Department of Agriculture. 2005. "Tomatoes: Shipping Point and Marketing Instructions." December 2005. https://www.ams.usda.gov/sites/default/files/media/Tomato_Inspection_Instructions%5B1%5D.pdf.

USDA, United States Department of Agriculture. 2018. "Vegetables 2017 Summary." February 2018. http://usda.mannlib.cornell.edu/usda/current/VegeSumm/VegeSumm-02-13-2018.pdf.

USDA, United States Department of Agriculture. n.d. "USDA Food Composition Databases." Accessed March 7, 2018. https://ndb.nal.usda.gov/ndb/search/list.

USDA ERS (United States Department of Agriculture, Economic Research Service). 2017. "2017 California Processing Tomato Report." 2017. https://www.nass.usda.gov/Statistics_by_State/California/Publications/Specialty_and_Other_Releases/Tomatoes/2017/201708ptom.pdf.

USDA ERS (United States Department of Agriculture, Economic Research Service). n.d. "Profit Margin Increases With Farm Size." Accessed September 18, 2018. https://www.ers.usda.gov/amber-waves/2015/januaryfebruary/profit-margin-increases-with-farm-size/.

USDA ERS (United States Department of Agriculture, Economic Research Service). 2018a. "USDA ERS—Farming and Farm Income." Accessed August 30, 2018. https://www.ers.usda.gov/data-products/ag-and-food-statistics-charting-the-essentials/farming-and-farm-income/.

USDA ERS (United States Department of Agriculture, Economic Research Service). 2018b. "USDA ERS—Ag and Food Sectors and the Economy." Accessed October 17, 2018. https://www.ers.usda.gov/data-products/ag-and-food-statistics-charting-the-essentials/ag-and-food-sectors-and-the-economy/.

US FDA, Food and Drug Administration. n.d. "Vitamins, Minerals." https://www.accessdata.fda.gov/scripts/InteractiveNutrition FactsLabel/factsheets/Vitamin_and_Mineral_Chart.pdf.

USDA, United States Department of Agriculture Foreign Agricultural Service. 2008. "The U.S. and World Tomato Situation." July 2008. http://usda.mannlib.cornell.edu/usda/current/tomwm/tomwm-07-30-2008.pdf.

Valero, C., C. Crisosto, D. Garner, E. Bowerman, and D. Slaughter. 2003. "Introducing Nondestructive Flesh Color and Firmness Sensors to the Tree Fruit Industry." In: *International Conference on Quality in Chains. An Integrated View on Fruit and Vegetable Quality* 604, 597–603.

Vegetable Facts. 2018. "Tomato History—Origin and History of Tomatoes." 2018. http://www.vegetablefacts.net/vegetable-history/history-of-tomatoes/.

Viviano, F. 2017. "*A Tiny Country Feeds the World*." Washington: National Geographic, September 2017.

Vrieling, A., D. W. Voskuil, J. M. Bonfrer, C. M. Korse, J. van Doorn, A. Cats, A. C. Depla et al. 2007. "Lycopene Supplementation Elevates Circulating Insulin-Like Growth Factor–Binding Protein-1 and -2 Concentrations in Persons at Greater Risk of Colorectal Cancer." *The American Journal of Clinical Nutrition* 86 (5): 1456–1462. https://doi.org/10.1093/ajcn/86.5.1456.

Wann, E. V. 1996. "Physical Characteristics of Mature Green and Ripe Tomato Fruit Tissue of Normal and Firm Genotypes." *Journal of the American Society for Horticultural Science* 121 (3): 380–383.

Washington State University Extension. 2018. "Tomato Varieties 2018|Skagit County|Washington State University." https://extension.wsu.edu/skagit/mg/plant-fair/tomato-varieties/.

Watanabe, T., M. Kioka, A. Fukushima, M. Morimoto, and H. Sawamura. 2014. "Biotin Content Table of Select Foods and Biotin Intake in Japanese". *Int J Anal Bio-Sci Vol* 2 (4).

Weaver, C. M. 2013. "Potassium and Health123." *Advances in Nutrition* 4 (3): 368S–377S. https://doi.org/10.3945/an.112.003533.

Wencel, D., T. Abel, and C. McDonagh. 2014. "Optical Chemical pH Sensors." *Analytical Chemistry* 86 (1): 15–29. https://doi.org/10.1021/ac4035168.

Wood, L. G., M. L. Garg, H. Powell, and P. G. Gibson. 2008. "Lycopene-Rich Treatments Modify Noneosinophilic Airway Inflammation in Asthma: Proof of Concept." *Free Radical Research* 42 (1): 94–102.

Yaegashi, Y., T. Onoda, K. Tanno, T. Kuribayashi, K. Sakata, and H. Orimo. 2008. "Association of Hip Fracture Incidence and Intake of Calcium, Magnesium, Vitamin D, and Vitamin K." *European Journal of Epidemiology* 23 (3): 219–225.

Yang, C.-C., Y.-L. Hsu, C.-C. Yang, and Y.-L. Hsu. 2010. "A Review of Accelerometry-Based Wearable Motion Detectors for Physical Activity Monitoring." *Sensors* 10 (8): 7772–7788. https://doi.org/10.3390/s100807772.

Yang, C.-C., M. S. Kim, P. Millner, K. Chao, B.-K. Cho, C. Mo, H. Lee, and D. E. Chan. 2014. "Development of Multispectral Imaging Algorithm for Detection of Frass on Mature Red Tomatoes." *Postharvest Biology and Technology* 93 (July): 1–8. https://doi.org/10.1016/j.postharvbio.2014.01.022.

Yearbook of Agriculture 1937. Department of Agriculture. http://archive.org/details/yoa1937.

Yuqing, M., C. Jianrong, and F. Keming. 2005. "New Technology for the Detection of pH." *Journal of Biochemical and Biophysical Methods* 63 (1): 1–9. https://doi.org/10.1016/j.jbbm.2005.02.001.

Zhang, H., K. Osada, M. Maebashi, M. Ito, M. Komai, and Y. Furukawa. 1996. "A High Biotin Diet Improves the Impaired Glucose Tolerance of Long-Term Spontaneously Hyperglycemic Rats with Non-Insulin-Dependent Diabetes Mellitus." *Journal of Nutritional Science and Vitaminology* 42 (6): 517–526. https://doi.org/10.3177/jnsv.42.517.

Zhang, M., C. D. Holman, and C. W. Binns. 2007. "Intake of Specific Carotenoids and the Risk of Epithelial Ovarian Cancer." *British Journal of Nutrition* 98 (1): 187–193.

Zhao, Y., M. L., Shi-Xuan Liu, and Q. Zhao. 2018. "Smart Hydrogel-Based Optical Fiber SPR Sensor for pH Measurements." *Sensors and Actuators B: Chemical* 261 (May): 226–32. https://doi.org/10.1016/j.snb.2018.01.120.

Zou, L., C. Ge, Z. Jane Wang, E. Cretu, and X. Li. 2017. "Novel Tactile Sensor Technology and Smart Tactile Sensing Systems: A Review." *Sensors* 17 (11).

Index

Printed in the United States
by Baker & Taylor Publisher Services